If there were no Shinkansen

Shuichiro Yamanouchi

If there were no Shinkansen

High-speed rail experience from its birth to today in Japan

 Springer

Shuichiro Yamanouchi
Toyko, Japan

ISBN 978-981-99-8889-1   ISBN 978-981-99-8890-7 (eBook)
https://doi.org/10.1007/978-981-99-8890-7

Translation from the Japanese language edition: "If there were no Shinkansen" by Shuichiro Yamanouchi, © 1999 2nd printing. Published by Tokyo Shinbun Publishing Bureau. All Rights Reserved.

© Japan Railway Technical Service 2024

This work is subject to copyright. All rights are solely and exclusively licensed by the Publisher, whether the whole or part of the material is concerned, specifically the rights of reprinting, reuse of illustrations, recitation, broadcasting, reproduction on microfilms or in any other physical way, and transmission or information storage and retrieval, electronic adaptation, computer software, or by similar or dissimilar methodology now known or hereafter developed.
The use of general descriptive names, registered names, trademarks, service marks, etc. in this publication does not imply, even in the absence of a specific statement, that such names are exempt from the relevant protective laws and regulations and therefore free for general use.
The publisher, the authors, and the editors are safe to assume that the advice and information in this book are believed to be true and accurate at the date of publication. Neither the publisher nor the authors or the editors give a warranty, expressed or implied, with respect to the material contained herein or for any errors or omissions that may have been made. The publisher remains neutral with regard to jurisdictional claims in published maps and institutional affiliations.

Cover design and illustration: Mariko Utsunomiya

This Springer imprint is published by the registered company Springer Nature Singapore Pte Ltd.
The registered company address is: 152 Beach Road, #21-01/04 Gateway East, Singapore 189721, Singapore

Paper in this product is recyclable.

Foreword

When I first met Shuichiro Yamanouchi, then Chairman of JR East, I immediately realized I was speaking with someone of immense presence, knowledge, and authority. Yet, he was kind and patient and answered my questions with care and consideration. The year was around 1998 and my queries concerned the Series 209 commuter train which had just been introduced on the Yamanote line around Tokyo. The strapline for this new model was *half the weight, half the cost, and half the lifetime.* I had just ridden on this vehicle and was immediately impressed by its clean and efficient lines. Yamanouchi expressed surprise that I knew something about the 209 and so started for me a much valued friendship.

This book is essentially an autobiography of Yamanouchi, and his life after graduating from Tokyo University and joining the national railway company JR in 1956, then subsequently with JR East following the privatization of JR in 1987.

Readers from other countries may be surprised learn that railway staff in Japan are highly trained on all the facets which go into running a railway and they usually stay with their company throughout their careers. Yamanouchi tells us of his extensive training, including driving a steam locomotive. During his career he espoused the philosophy of the railway as a system with people at its heart.

However, in the 1950s and 60s there was considerable strife among the railway workers. I well recall the so-called *spring offensive* which resulted in strikes as pay and conditions for the following financial year were vigorously negotiated. Yamanouchi describes how this deterioration continued that overmanning was rife and the financial position of JR declined in an unsustainable way. This prompted the privatization of the system in a manner which differed from that employed in many other countries. Vertical integration of trains, track, and operations, was retained and the system split on a regional basis, together with a national rail freight company. On the main island of Honshu, companies West, Central, and East were formed, and the islands of Kyushu, Shikoku, and Hokkaido had their own companies. The main island companies prospered, industrial relations were improved, while manning levels were reduced, but the island companies, lacking sufficient population density, together with ageing and further depopulation, struggled, and closures were made, a situation which continues to this day.

JR Central has been a success based on the profitable Tokaido *Shinkansen*, whilst JR East captured the huge commuting market in and out of Tokyo, making it one of the largest railway companies in the world. It's statistics are staggering; in 2019 JR East carried 18 million passenger every day of which 1.6 million passed through the busiest station, Shinjuku.

The title of Yamanouchi's book indicates its main thrust, the building, and evolution of the Japanese shinkansen network and its role in the economic success of the country. Particularly interesting are the debates and opposition to the construction of the Tokaido Shinkansen which way back in 1964 opened as the world's first high-speed line, linking Tokyo with Osaka. This is, of course, the most densely used high-speed railway in the world. But its birth was far from easy. JR East's high-speed lines are not as well used but have played a key role in shrinking the country north and west of Tokyo. Readers will find the details of the evolution and speed-up of high-speed rail in Japan particularly interesting, as are the details of the complexities of financing such mega projects.

After his retirement from JR East in 2000, Yamanouchi became involved in Japan's space programme, becoming President of the Japan Aerospace Exploration Agency (JAXA). He died, too young, from heart failure in 2008, aged 75. The editors of the present book have done a great service to the international community, by making available Yamanouchi's original manuscript to a wide audience. This volume is a tribute to the life of a remarkable railway man, a leader in Japan, and a fervent supporter of what he called the *important social capital of the shinkansen.*

May 2024   Roderick A Smith
                                                                  Imperial College London
                                                                              London, UK

Editing Notes

Introduction

The title of this book, *If there were no Shinkansen*, is a very sensational one. But the answer is simple: Otherwise, Japan would not be the prosperous country it is today, and the world's railways would not have attained the high speeds they have.

This book is written not only for railway professionals but also for the general reader interested in railways. For this reason, it was initially published in a hardcover edition by the Tokyo Shinbunsha and later in a paperback edition by the Asahi Shinbunsha. Although the number of copies is not disclosed, many copies of the book have been printed. The author, Mr. Yamanouchi, was an engineer with long experience working for Japanese National Railways in the field of train operation and safety. He also had a managerial perspective through working on the reform of Japanese National Railways and serving as Chairman of East Japan Railway Company. Thus, I am convinced that this is the best book to introduce Japanese railways to people overseas, as it is written from a broad perspective on both technical and management perspectives.

One of the themes of this book is the mutually stimulating progress of the world's high-speed railways. The book also empirically demonstrates that new technologies emerge through various setbacks. It takes a lot of courage to move forward into an unknown new world. This book will be a great reference for those who are looking for a new challenge in the world of railways to see how their predecessors gained a foothold in the new world.

I have had long experience in assisting developing countries through ODA-related overseas railway technical cooperation. Through these experiences, I have learned that the knowledge of the railway professionals in these countries about railways is largely from a European viewpoint since few books on Japanese railways have been published in English. If they are going to use Japanese railway technology, we want them to learn about Japanese railways and fully utilize the technology that is available to them. In addition, the world's railways are not in competition with each other regarding railway operation business. Only through the mutual exchange of experiences can each railway develop. These two thoughts inspired me to launch a project to publish a book about Japanese railways internationally. However, an English translation is required to publish a book in English. I did not know how many readers would buy a book about Japanese railways. Perhaps the first challenge is to accept a loss. Fortunately, this book has a private edition for distribution at the UIC. Although this book was originally written for Japanese readers, it can unexpectedly provide an introduction to Japanese railways for foreign readers, as it describes various scenes experienced by the author at JNR during those turbulent times. Mr. Yamanouchi must have been aware of this and chose it with the expectation that UIC members would welcome it. The book contains many stories of failure, and some people were anxious that it would disgrace Japan to the rest of the world. However, the exchange of experience is only meaningful if it includes failure. For this reason, we have selected this book as the first in a project to publish books on Japanese railways from an international publisher. Note that although this book was written by Mr. Yamanouchi more than 20 years ago, it deals with historical facts and still provides fresh inspiration to those involved in the railway industry. However, since it was written more than 20 years ago, we have decided to provide notes and recent information based on assumptions concerning what Mr. Yamanouchi would write in this book now.

We hope that many people will become interested in Japan's railways through this book.

Acknowledgements

The English version of this book was published through the generosity of Mr. Hideki Yamanouchi, Ms. Marie Nakajima, and Mr. Nobuaki Nakajima, family members of the late Mr. Yamanouchi, the author of the book.

The Tokyo Shinbunsha Publishing Bureau, which published the Japanese edition, has kindly agreed to publish the English edition. We would like to express our sincere gratitude to them.

We would also like to express our gratitude to the many people and organizations listed below who have supported us in various ways.

Dr. Roderik A Smith, Professor Emeritus, Imperial College London, President of the Japan Railway Society; Dr. Takashi Yamanouchi, Professor Emeritus, Institute of Polar Research; Dr. Anthony Robins, Professor, Aichi University of Education and Coordinator in Japan of the Japanese Railway Society of the United Kingdom, and Ms. Alexandra Lefebvre of International Union of Railways (UIC).

The cover of this book indicates that it is not just about the *Shinkansen* but also includes a diverse range of railway topics. The design, including Mr. Yamanouchi's silhouette, was created by Ms. Mariko Utsunomiya, a JR East/Japan International Consultants for Transport officer with design talent.

The Japanese version of this book contained monochrome photographs, but a policy was established to replace these with color photographs as much as possible in the English version. The following people and organizations have contributed to this initiative.

Professional railway photographer; Mr. Masatoki Minami, Dr. Anthony Robins, and Keio University Railway Research Association alumni: Mr. Katsuji Iwasa, Mr. Tadashi Sumita, Mr. Keiji Musha, and Mr. Kuniaki Mori.

JR East, Japan Railway Construction, Transport and Technology Agency (JRTT), Kotsu Kensetsu, The Railway Museum, Kotsu Shinbunsha, Tokyo Shinbunsha, Tokyo Station Gallery, and Gala Yuzawa Corporation.

This book is based on an existing English translation for private use and was voluntarily brushed up by Professor Anthony. Furthermore, I am indebted to two of my senior colleagues, Mr. Katsuji Iwasa and Mr. Tadamasa Nagai, for their great help in preparing the footnotes already mentioned and scrutinizing the author's Japanese to convey precisely what he intended to say even in English. All three of us, including myself, have worked under Mr. Yamanouchi at JNR in the same field and know exactly what he means. This good team is what made this book possible.

Many more JR Group officials and retirees assisted in publishing this book. The following is a list of those not already indicated.

Dr.Tatsuhiko Suga, Dr.Tetsuo Shimomae, Mr.Takao Kubo, Mr.Kazutoshi Watabe, Mr.Toru Fukushima, Mr.Hiroyuki Nakamura, Mr. Yoshihiro Akiyama, Dr. Fumio Kurosaki, Mr. Fumihiro Araki, Mr. Hiroshi Komatsu, Mr. Hitoshi Saimyo, Mr. Shunzo Miyake, Mr. Takahiro Kikuchi, and Mr. Shunsuke Takagi.

At last, we would like to express our deepest gratitude to Springer for accepting this book for publication. We would also like to thank Mr. Smith Chae, Editor of Springer Nature Korea Limited, for his excellent management in preparing and publishing this book, and Ms. Saranya Devi Balasubramanian and Ms. Sangeetha Ganesan, Project Coordinators of Springer Nature India, for their direct assistance.

May 2024   Tetsuro Aikawa
                                                                                        Representative of Publishing Project
                                                                                       for the English version of 'If there were
                                                                                                                    no Shinkansen'

Contents

About the Author

Late Shuichiro Yamanouchi

Photo provided by his family

Born in 1933
1956 Graduated from the Faculty of Engineering, The University of Tokyo
1956 Joined Japanese National Railways
1981 Director General, Tokyo Northern Railway Administration Bureau
1982 Director General, Head Office Operations Bureau
1985 Managing Director of Japanese National Railways
1987 Deputy President of East Japan Railway Company

1996 Chairman of East Japan Railway Company
Appointed to Vice Chairman of International Union of Railways (UIC)
2000 Chairman, National Space Development Agency of Japan
2008 Passed away.

Chapter 1
The Great Railway Nation of Japan

As Japan is known as a 'railway country,' railways are well developed in Japan and have considerably more passengers than in other countries. However, most of them were concentrated in metropolitan areas, mainly Tokyo, Nagoya, and Osaka, where the tracks were meter gauge. Why was a standard gauge Shinkansen born in such a Japan?

At the time, a number of steam locomotives were still in service in Japan. At the same time, electrification work was completed on the conventional line between Tokyo and Osaka, and the high-speed EMU train *Kodama* began to run; this was when the momentum for high-speed transit between major cities was beginning to build up.

1.1 Half of the World's Rail Passengers Are in Japan

About four years ago,[1] I found an article describing Japanese railways in a French railway magazine. Reading through it, I was surprised to see the statement: "Half the people who use rail transportation in the world on any given day are Japanese." I quickly looked into the matter and found that about 160 million people in the world use rail transportation on any given day. And indeed, 62 million of them are in Japan.[2]

When we talk about the whole world, we are, of course, including such countries as China, India, and Russia, and Japanese passengers account for not 50% of the total, but 40%. These figures include people who transfer among the JR, the private railways, and the subways, which are all owned by different companies and are

---

[1] From the time of publication of this English version, it was around 30 years ago.

[2] According to JR East's Annual Report and other sources, at least 200 million people use railways daily in the world today, of which the number of Japanese rail users has increased to 69 million.

© Japan Railway Technical Service 2024
S. Yamanouchi, *If there were no Shinkansen*,
https://doi.org/10.1007/978-981-99-8890-7_1

therefore counted two or three times, so the actual number is somewhat smaller. Even so, Japan is a great railway nation.

Here, I may need to provide some explanation.

Fifty-five million, or 88% of the passengers on Japan's railways, are actually in the three metropolitan areas centered on Tokyo, Nagoya, and Osaka. Most of these are people who buy monthly passes and commute to work or to school.

Yet China's railways carry almost no passengers of this type. It is not necessarily the case that since Beijing and Shanghai have subways, they don't have railways for urban transportation, but the majority of the trains of the Chinese National Railway, which owns the tracks in that vast nation, are medium- and long-distance trains similar to Japan's express trains. There are no tracks like those of the Yamanote Line or the Chuo Line that are used for urban transportation, and the daily passenger load of the Chinese National Railway is a mere 2. 6 million people.[3] Bicycles and automobiles, particularly minibuses, play the main role in urban transportation, and their routes cover China's cities in a net-like pattern.

China thus has a relatively small amount of railway tracks in proportion to its huge land area. Japan has 27,000 km of rail lines, while China has only 60,000 km. China is building 1,000 km of new rail lines each year, but that is still a small amount for a land mass twenty-five times the size of Japan.[4]

It is inappropriate to compare the scale of railways merely by the number of passengers. That is because the same person may ride from Tokyo Station to Yurakucho Station, which is also within the city, or from Tokyo Station to Shin-Osaka Station, several hours away. The unit usually used for comparing the passenger loads for various means of transportation is 'passenger kilometers.' This represents the distance that the passengers travel by train or airplane. When we calculate these numbers, we find that Japan has the largest passenger load in the world.

Let's see how that compares to the situation with European railways.

The passenger load of JR East is larger than those of France and Germany combined. The combined passenger load of the six JR companies corresponding to the old Japanese National Railways is greater than that of ten nations of the European Union. If we add Japan's privately owned railways, the total passenger load is greater than in all of Europe, excluding the former Soviet Union. This is despite the fact that Japan's land area is only one-sixtieth that of Europe.

Until just recently, JR East was the world's largest transport company. I say 'was' because the recent depreciation of the yen has allowed the American package delivery service United Parcel Service (UPS) to take first place. Even so, JR East is the world's

---

[3] As the author states, China's railways used to be mostly intercity lines operated by China National Railways. However, in 1969, the first subway system opened in Beijing. In 1978, the modernization policy based on the reform and open-door policy led to the rapid construction of urban railways such as subways, trams, and monorails in significant cities. As a result, as of 2022, urban railways operate in about 40 cities, covering about 250 routes and 8,700 operating kilometers.

[4] In China, the operating kilometers of high-speed and urban railways developed over the next two decades have increased to 146,000 km, approximately 2.4 times that of that time. In particular, the high-speed railway network covers approximately 38,000 km throughout China.

Fig 1.1 Shinjuku Station, shown in 2022, which serves more passengers than any other station in the world. *Photo* provided by Tetsuro Aikawa

largest railway company, larger than German Railways, and its sales are far greater than those of the world's largest airlines, American and United.

Looking at the items related to railways in the Guinness Book of World Records, I found that Japan is mentioned twice.

The first record is for the railway passing through the lowest elevation, and this honor went to the Seikan Tunnel, which runs 240 m under the straits between the islands of Honshu and Hokkaido.

The other record is for the world's busiest railway, and it informs the reader that JR East carries an average of 16.3 million passengers per day.[5]

JR East moved its headquarters from the Marunouchi area of central Tokyo to Shinjuku Ward in the fall of 1997. In addition to JR East, ten rail lines run in and out of Shinjuku Station, including the Odakyu, Keio Teito, and Seibu private commuter lines and subway lines. 3.2 million people get on and off trains here in a single day, and Shinjuku Station alone has more passengers than all of China. There is probably no other station like it in the world (Fig. 1.1).

Furthermore, Japan has 190 railway companies. There is probably no other nation like it in the world.

---

[5] As of 2019, JR East had an average of 17.8 million passengers per day.

Viewed in terms of passenger load, the private railways, including the subways, carry 50% more passengers than the JR Group. The major private railway companies have also branched out beyond their core business of the railway to venture into hotels, department stores, real estate, and recreational facilities and have become some of Japan's foremost corporate groups.

Japan is also known as the 'kingdom of the private railways.'

1.2   If There Were no *Shinkansen*[6]

Recently, a French railway executive visiting Japan made an interesting comment.

> When I stay at a hotel in Europe, I can often see train tracks from my window, but hardly any trains come by. When I came to Japan and stayed in a hotel in Kyoto, there seemed to be a *Shinkansen* running past whenever I looked out the window.

At its busiest times, the Tokaido *Shinkansen*, which runs from Tokyo to Shin-Osaka, has as many as eleven trains running every hour.[7] If you take the round trips into consideration, twenty or more trains are running at any one time, so they come through Kyoto every three minutes. No wonder the Frenchman made that remark.

I may safely say that there is no railway like this anywhere in the world. In any case, the total passenger kilometers of the Tokaido *Shinkansen* are equivalent to 70% of the passenger load of the entire French National Railway system.

The line with the highest passenger load on the TGV, the French counterpart to the *Shinkansen*, is the line running between Paris and Lyon. It carries 48,000 passengers per day, about 1/7 the number for the Tokaido *Shinkansen* and about the same as the Joetsu *Shinkansen*, which runs from Tokyo to Niigata.[8]

Despite this, the Paris-Lyon TGV seems to be a money maker for the French National Railway. President Louis Gallois, who visited Japan in 1997, told me, "If we could keep just the TGV between Paris and Lyon and get rid of all our other rail lines, that would be the best thing that could happen."

What would have happened if Japanese National Railways had not created the Tokaido *Shinkansen*?

The *Kodama*, the fastest limited express train before the *Shinkansen* was built, took six-and-a-half hours to run between Tokyo and Osaka. The old Tokaido Line, with its many sharp curves and turnouts, made it difficult to increase speeds any further.

---

[6] *Shinkansen* trains are also called bullet trains, but in this book, they are referred to as *Shinkansen* (which means new truck line in Japanese)

[7] Currently, the maximum number is 12 per hour.

[8] As for the route between Paris and Lyon, SNCF began operating the low-cost TGV "OUIGO" in 2013 in order to strengthen its competitiveness not only with airplanes but also with automobiles. In addition, Trenitalia (Italian Railways) newly entered this section in December 2021, and although the details are not clear since SNCF does not currently publish the number of passengers transported by section, it is believed that the number of passengers on the 'Paris-Lyon' route has grown considerably since then due to these factors.

Fig 1.2 A crowded Tokyo Station platform for the retirement ceremony of the Series 100 *Shinkansen.* The Tokaido *Shinkansen* is undergoing a generational shift with the original Series 0, Series 100, Series 300, Series 700, and Series N700. *Photo* provided by Kotsu Shinbunsha

At that time, airfares were still far more expensive than train fares, so a lot of passengers still rode the trains, but now that there isn't such a great difference between the two fares, there would undoubtedly be hardly any passengers taking the six-and-a-half-hour trip by train. And, of course, the Sanyo *Shinkansen* to Hakata in Kyushu, the Tohoku *Shinkansen* to Morioka, and the Joetsu *Shinkansen* to Niigata would never have been built.

If we continue thinking along those lines, we can imagine that if we had not made the bold decision to build the *Shinkansen* about forty years ago, the Japanese railway system might now consist almost entirely of money-losing local lines, except for lines in large cities such as Tokyo and Osaka. Or it might have disappeared altogether. That one decision saved Japan's railways (Fig. 1.2).

But that's not all. The success of Japan's *Shinkansen* provided a shock and a stimulus to European countries, which had well-developed rail systems, and it gave rise to France's TGV and Germany's ICE systems. If Japan had not built the *Shinkansen*, then Europe's high-speed railways would never have come into being, and railway passenger service would undoubtedly have gone into decline.

Furthermore, in Asian countries such as South Korea, as well as China and Taiwan, plans are underway for high-speed rail services like the *Shinkansen.*[9] The decision to build the *Shinkansen* was one that determined the future of the railway industry worldwide.

---

[9] High-speed rail is now in operation in these countries.

If the *Shinkansen* had never been built, would it have been possible to move the same number of passengers by airplane? The Tokaido *Shinkansen* is mostly made up of sixteen-car trains running at intervals of five minutes.

I tried calculating the number of airplanes that would be needed if we were to experiment with transporting the Tokaido *Shinkansen* passengers by air.

I suppose that the *Shinkansen* passengers who were traveling short distances would switch to car travel, so I assumed that those passengers traveling more than 300 km would travel by air.

I came up with the answer that we would need about a hundred jumbo jets. The passenger load between Tokyo and Osaka would increase by a number corresponding to half the passenger load of all of Japan's domestic airlines. The route between Tokyo and Sapporo is the air route with the greatest number of passengers in the world,[10] at about 20,000 people per day, but without the *Shinkansen*, the domestic carriers would have to transport nine times that number of people. That would probably be quite impossible.

Then what would happen if we moved all these people by bus instead of by *Shinkansen*?

Shinkansen trains are made up of sixteen cars and can carry 1,300 passengers, and at the busiest times, eleven[11] trains leave every hour. To transport the same number of people in 40-passenger buses, we would have to run buses at intervals of every ten seconds. Actually, we would probably end up using both buses and airplanes, but even with that, it would be impossible to transport the same number of passengers as the *Shinkansen*.

The February 21, 1998 issue of the famous British economics magazine *The Economist* contained an article, 'A New age to Railways,' which stated, "If the hundreds of millions who travel on these *Shinkansen* express lines each year switched to car travel, there would be at least 1,800 extra deaths and 10,000 serious injuries each year."

If the *Shinkansen* had not been built, Japan's current economic growth might have been impossible. This one decision may have controlled the future of the nation's economy.

1.3   How Did the *Shinkansen* Come to Be?

Formal planning for the construction of the Tokaido *Shinkansen* began in 1956. In May of that year, Japanese National Railways set up the Tokaido Line Augmentation Investigative Conference to begin looking into means of enhancing the transportation capability of the Tokaido main line, which had already reached the limits of its capacity.

---

[10] Currently, the busiest air route in the world is Seoul-Jeju with 39,700 passengers per day (IATA, 2018)

[11] Now it is twelve.

At the outset of the conference, Shinji Sogo, president of JNR, stated, "In my opinion, if we are going to enhance the Tokaido Line, the answer is standard gauge tracks. We must resize the gauge, increase speeds, and transport larger amounts of passengers and cargo."

At that time, 120 trains per day ran each way on the Tokaido Line, near the limit for a double-track line. The roads, too, were still in a terrible state.

In fact, the situation with roads was so bad that the Watkins Economic Fact-Finding Group from the United States, which visited Japan in the same year, pointed out, "Japan has designated places where it plans to build highways, but it doesn't have the actual highways yet."

The year 1956 was a great turning point for Japanese society. That was the year when the famous words, "We will soon move out of the postwar period," appeared in the Economic White Paper, as well as the year when Japan joined the United Nations.

That was also the year when the Japan Highway Public Corporation was launched. It may be said that this was the era when setting up a transportation network became the central focus of national development plans. In the following year, the Law on Constructing Automobile Roads Across the Country for National Development was enacted, and in 1958, construction began on the Meishin Expressway between Nagoya and Kobe.

On the world scene, 1956 was the year of the Suez crisis and the Hungarian uprising. Incidentally, it was in the previous year that the Liberal Democratic Party and the Social Democratic Party of Japan were launched, and the so-called fifty-five-year system was put in place.[12]

In the fall of that year, the electrification of the entire length of the Tokaido Line was finally completed, and it became possible for people to travel without being bothered by smoke. The fastest train at that time, the JNR's flagship limited express train, the *Tsubame*, shortened the time required to travel between Tokyo and Osaka from eight hours to seven-and-a-half hours for the first reduction in twenty-two years (Fig. 1.3).

It was also in that year that experiments began on changing the power supply systems to alternating current (AC), one of the essential technologies for creating the *Shinkansen*.

It was precisely that year that I began working for Japanese National Railways (JNR), but in Europe and the United States, the sun seemed to be setting on the railways as we moved into the age of aviation and automobiles.

The university professor who advised me on finding a job commented, "Isn't it a bit late in the day to be getting into the railway business?".

Most of my friends headed for large companies in growing industries such as heavy electrical equipment, automobiles, and shipbuilding. Textile companies were another popular choice. SONY didn't yet exist, although people had heard of its

---

[12] The '55-year regime' was the system under which the ruling Liberal Democratic Party (LDP) maintained a continuous majority in Japan while the combined opposition parties, including the Japan Socialist Party (JSDP) and other non-LDP parties, held more than one-third of the seats in the Diet, thus preventing LDP from holding two-thirds of the seats, which is a requirement for constitutional reform.

Fig 1.3   Limited express *Tsubame* with an electric locomotive hauling newly green repainted coaches introduced after the electrification completion of the Tokaido Line. At Nagoya staiton Nov 1956. *Photo* provided by Katsuji Iwasa

precursor, Tokyo Telecommunications, but it still wasn't a very large company, and Matsushita Electric[13] was still just a manufacturer of household electrical products, with no one aspiring to a position there. Even Toyota was nothing more than a plant in Aichi Prefecture, and no one tried to get a job there.

At that time, JNR was still a company that one could be extremely proud of. It enjoyed much larger sales than NTT and Toyota, and it made a profit. It did not issue individual private job offers to students. Instead, applicants had to pass through a full day of written exams, group discussions, and individual interviews in order to join the company. No one could have imagined that the world's fastest train would come into being a mere eight years later and that thirty-one years after that, JNR would be privatized after having descended into bankruptcy.

To be frank, I met with a series of disappointments after joining the company.

Soon after my generation entered university, we participated in part of the May Day incident,[14] and since we held strongly negative feelings about war and mistrusted the national system, the old-fashioned, overly formal, and rule-ridden corporate culture didn't agree with us.

---

[13] Now it has been changed to "Panasonic.".

[14] On May 1, 1952, the 23rd May Day, demonstrators opposing the San Francisco Peace Treaty and the Japan-U.S. Security Treaty clashed with police in Tokyo, resulting in two deaths and more than a thousand people seriously injured, in what is also known as the 'Bloody May Day incident.'.

The managers' lectures were nearly devoid of content and consequently boring, and we reacted against the corporate culture that looked down on technicians. This was an era when technology was mostly centered on steam locomotives, so it didn't arouse any interest at all. And that was the time when the idea of building the *Shinkansen* was born, and we were able to have big dreams.

The idea of the *Shinkansen* arose out of one dream and one reality. The dream was that of President Shinji Sogo and other executives from the pre-war era of building a high-speed railway that would be equal to any in Europe or the United States. The reality was that the Tokaido Line had already reached the limits of its capacity.

It was only natural that the dreamers would have insisted on building new tracks of the gauge that was standard in Europe and the United States (a width of 1,435 mm between rails) and that the realists insisted on building quadruple meter gauge tracks, starting with the sections that ran the most trains and continuing on down in order. They wanted to increase the number of trains running as soon as possible.

For this reason, the Tokaido Line Augmentation Investigative Conference began deliberations to compare a proposal for parallel meter gauge tracks, a proposal for dedicated meter gauge tracks, and a proposal for new, standard gauge tracks. However, it was nearly impossible for even JNR to bring about such a large project on its own. There would have to be coordination with the plans for the expressways. Plans for the construction of the expressways had made substantial progress, and work began on the Meishin Expressway, Japan's first expressway, in 1958.

In response to requests from JNR, the government set up the Japanese National Railways' Trunk Line Investigative Conference in August 1957, and in July of the following year, the Investigative Conference presented its findings to the Minister of Transport. The report emphasized that JNR needed to construct a new line as soon as possible. It asked that the new system be "the most modern means of transportation at the world's highest technological level" and specified that "a dedicated standard gauge line would be appropriate."

During these deliberations, there were not only disputes within JNR between the advocates of a new standard gauge trunk line and the advocates of quadruple meter gauge tracks but also fierce disputes with the advocates of highways and airplanes.

The mass media criticized the idea of investing massive amounts of money in railways, which were in decline in Europe and the United States, and building a new trunk line. They even referred to it as one of the three great follies of the world, along with the Great Wall of China and the battleship Yamato.

The construction of a new trunk line was approved because it would have much greater transport capacity than a highway, would yield sufficient profit, and because the necessary funds could be procured without raising fares. For that reason, the estimate for total construction costs was low, at 197.2 billion yen. That estimate later led to major problems.

JNR set up an internal Trunk Line Investigative Office parallel to the government's Trunk Line Investigative Conference and began looking at specific routes for the line and technological standards.

Previously, in May of that same year, the Railway Technical Research Institute held a lecture meeting to commemorate the fiftieth anniversary of its founding at

Fig 1.4   Dr. Takeshi Shinohara, the General Manager of the Railway Technical Research Institute, at its fiftieth anniversary lecture meeting, May 30, 1958, at Tokyo's Yamaha Hall. *Photo* provided by Kotsu Shinbunsha

Yamaha Hall in the Ginza area of Tokyo. It was announced that a railway with trains running at speeds of up to 250 km/h and linking Tokyo and Osaka in three hours was mostly technologically feasible (Fig. 1.4).

Preparations were moving along at the pace desired by the dreamers, but the realists also had some reasonable points to make.

In 1957, JNR started making its first Five-Year Plan with total construction costs of 597 billion yen, but the plan centered on the replacement of worn-out, facilities improvement of safety measures, and enhancement of transport capacity. During the war, both railcars and facilities had been used under harsh conditions, and even afterward, it had not been possible to repair or replace enough of them, so the JNR's facilities had fallen into a pretty terrible situation.

There were also many major accidents. In 1954, a ferry between Aomori and Hakodate, the Toya-maru, sank in a typhoon, with the loss of 1,430 lives, and in the following year, 168 people died when the Shiun-maru, a ferry on the Uko Line, sank.

The railways, too, had their share of accidents. In 1956, forty people died in a train collision on the Sangu Line.

The commuter lines were horribly crowded, and the long-distance trains didn't run often enough. There were mountains of projects on hold: building new rolling stock, converting the single-track sections to double tracks and electrification.

At this time, the entire length of the Tohoku main line north of Utsunomiya was only a single track, and there were still no super express trains. The ordinary expresses

Fig 1.5 Handing over the tablet token. *Photo* provided by the Railway Museum

made only five round trips per day, and the entire section from Tokyo's Ueno Station to Aomori was served by steam locomotives. The trip took fourteen hours.

The Chuo Line running west out of Tokyo was also single-track beyond Takao, and while working as a driver's assistant, I was assigned the job of handing over and receiving the tablet token that served as a voucher for the block system, between

Takao and Kofu. This was much less easy than you would expect. I was told that taking the tablet token down from the pillar where it was attached, to my arm was dangerous, so I took it in the palm of my left hand, but in cases where the train did not stop at the station, even if I properly located the tablet token's position and tried to take it, I often couldn't get the timing right (Fig. 1.5).

Just about that time, I had the occasion to visit JNR's main office, probably because that was when a conference about the plans for the new trunk line was being held. The director of the Train Schedule Department, Train Operations Bureau, came back into the room and indignantly slammed his documents down on the desk. "It's no good! They just won't understand!" he fumed.

Since he wanted to increase the number of trains on the Tokaido main line as soon as possible, he did not want to build a new high-speed trunk line. Instead, he insisted on building quadruple meter gauge tracks parallel to the existing ones, but his proposal had not been accepted.

This mood spread throughout JNR. The New Line Investigative Office was called 'the Daydream Corps.' This probably reflected the difference between the leaders who had dreams and visions and the practical types, who, despite their admirable qualities, could not separate themselves from the current reality.

However, it was also true that if the leaders' dreams were mistaken, they would lead to the ruin of the business.

In its report, the New Line Investigative Conference formally decided that the new trunk line would be standard gauge, and on April 13, 1959, the *Shinkansen* project received formal authorization. A groundbreaking ceremony was held at the east entrance of the Shin-Tanna tunnel on April 20.

1.4   The Course of Development of Japan's Railways

Why had Japan's railways chosen to lay meter gauge tracks? A widespread theory maintains that the Englishmen who acted as advisors when Japan's railways were constructed recommended a meter gauge system of the type specified for their colonies because they thought that it would be sufficient for a backward country like Japan. Yet the facts of the matter are not necessarily clear.

When it constructed the railways, the Japanese government hired an English engineer named White, and he proposed a meter gauge railway three feet, six inches wide (1,067 mm) to Edmund Morel, a master builder responsible for the construction technology of the rail system. It is true that Morel abided by this opinion, but we do not know what the Japanese government thought of it.

Still, the Japanese government did not necessarily assent to whatever the British said throughout the project. In fact, their opinions were at odds on many points. It is possible, therefore, that the Japanese government determined that meter gauge tracks were the best option for quickly constructing a railway network for their small country.

But as early as 1888, there was a movement to switch over to standard gauge tracks. The army promoted the idea of refurbishing the tracks, claiming that the meter gauge tracks didn't allow enough transport capacity and that Japan should follow the example of Western countries and change over to standard gauge.

The Sino-Japanese War provided a further spur to this movement, and in 1893, the Lower House of the Diet approved a proposal to convert the tracks to standard gauge. The Communications Ministry appointed 'Track Investigation Committee members' to carry out studies, but in the end, the discussions gradually lost their momentum. The Army strongly urged conversion of the track gauge, but the Communications Ministry, which had jurisdiction over the railway, was relatively cool to the idea of conversion.

Since huge amounts of funds would be needed to convert the track gauge from meter gauge to standard gauge, it was impossible from a financial point of view, and even if only the main lines of major military importance were converted, having two track gauges would indeed make things more inconvenient.

Discussions about converting the track gauge gradually died down in the late 1880s and early 1890s, and the attention of the government and the military shifted from converting the gauge to discussions of nationalizing the railways.

This would be a good place to describe the structure and organization of Japan's railways.

In 1872, when Japan's first railway began operations, the Railway Office of the Ministry of Public Works was in charge of constructing and operating railways. The Railway Office later became the Railway Bureau, and in 1890, the entire length of the Tokaido Line was opened. The following year, the Railway Bureau became the Railway Agency, an independent organization under the control of the Ministry of Internal Affairs.

Then, in 1892, control of the Railway Agency was shifted from the Ministry of Internal Affairs to the Ministry of Communications, and a new Railway Conference was founded as an advisory body for railway policy. Even though the Railway Conference was called an advisory body, it was an organization with a great deal of power, having authority to determine the order in which construction on new railways would start, determine offerings of public bonds, study budgets for construction work, and approve or deny licenses for private railways.

The 'Railway Construction Law' was enacted at about the same time. This law determined the basic plans for the construction of new lines and the order in which construction would begin, and in later years, it became the motivation for the construction of unnecessary, money-losing local lines.

Soon after that, the Railway Agency went from being an independent government agency to being the Railway Bureau, one of three bureaus in the Ministry of Communications, along with the Communication Bureau and the Shipping Administration Bureau. This ministry was the government agency in charge of communications, and land and marine transportation.

However, as the railway network developed, it became difficult for one bureau to be in charge of both the construction and operation of railways. In 1903, the business divisions were spun off as a separate bureau, giving rise to the 'Railway Works Bureau.'

A number of today's rail lines, including the Tohoku Line to northern Japan, the Sanyo Line in western Honshu, and the Chuo Line, were built by private railways during that era, and eventually, the private routes were three times larger than the government-managed railways, with many of them offering speed and service superior to that of the government railways.

At the end of the Japanese-Russo War, both the government and the economic leaders, with strong support from the military sector, insisted that the railways should be nationalized as a basis for economic expansion. In 1906, it was decided that the railways would be nationalized, and within a mere two years, the government had bought up 4,800 km of routes from 17 railway companies, including Nippon Railway (the Tohoku Line and others), the Kobu Railway (today's Chuo Line), and the Sanyo Railway (the Sanyo Line). At the same time, 48,000 employees and 25,000 pieces of rolling stock were transferred from the railway companies to the new National Railway.

At the same time that the railways were nationalized, the Railway Works Bureau was abolished, and the Imperial Railway Agency came into being. But one year later, it was detached from the Ministry of Communications and became the 'Railway Authority,' directly under the control of the Cabinet. We might call this the actual birth of what later became 'Japanese National Railways.'

Shinpei Goto, former president of the South Manchuria Railway, assumed the post of the first president.

The nationalization of the railways became the opportunity to construct a central station in Tokyo, and in 1914, the familiar red brick building known as Tokyo Station was constructed according to a design by Tokyo Imperial University Professor Kingo Tatsuno.

With its completion, Tokyo Station became the new terminus of the Tokaido Line, a function formerly held by Shinbashi Station. At the same time, trains began running between Tokyo Station and Yokohama on the present-day Keihin Tohoku Line.

The Chuo Line was extended to Tokyo Station in 1919, and the Yamanote Line began its present-day operation as a circular commuter line in 1925.

With the nationalization of the railways, the railway divisions making up the Railway Authority gained their independence from the Ministry of Communications and became a full-fledged organization. In those days, when the railways had a monopoly on land-based transportation, the Railway Authority became a powerful economic agency in charge of on-site operations, with authority over not only the construction and operation of railways but also the licensing and supervision of the private railways and supervision of the Korea Railway and the South Manchuria Railway. Yet the railway officials were not at all satisfied with this.

That is because the president of the Railway Authority was not a minister of state, and could neither attend Cabinet meetings nor determine policies freely. A vigorous movement for change paid off, and in 1920, the Railway Authority had its wish fulfilled when it was promoted to Cabinet level and renamed the Ministry of Railways. It was about this time that the first steam locomotive really produced in

Japan, the C51, came into being, and Japan's railways gradually moved out of an era of importing and imitating Western technology and into an era of inventing their own.

Afterward, during the move toward administrative simplification during World War II, the Ministry of Railways was merged with the Ministry of Communications to form the Ministry of Transport and Communications, but that period didn't last long, and shortly before the end of the war, the communications divisions were once again split off, and the Ministry of Transport was formed.

After the end of the war, in 1949, the operational divisions of the railways were spun off from the Ministry of Transport, and a public corporation known as 'Japanese National Railways (JNR)' was inaugurated. This happened under orders from the Allied Armed Forces, and labor problems lay behind the move. The Occupation troops stationed in Japan after the war were promoting the democratization of the country's social systems, and one of their policies was to actively foster labor unions.

The new Constitution enacted in 1946 recognized such basic labor laws as workers' rights of organization, rights of collective bargaining, and the right to strike. National public employees were no exception to this law, but since the public employees, especially those associated with the national railways, were making plans for a general strike on February 1, 1947, in order to demand year-end lump-sum payments and a minimum wage system, the Supreme Allied Command feared that there would be social unrest. They ordered a halt to the strike and asked that the National Public Service Law be revised so that the right of public employees to collective bargaining and to strike would no longer be acknowledged.

JNR was set up as a 'public corporation' whose employees had the right to organize and the right to collective bargaining but not the right to strike. However, despite their supposed lack of a right to strike, the railway employees struck often, and labor-management relations gradually worsened.

A public corporation is likely to be subject to political intervention, and with its weak sense of responsibility and autonomy, the railway was headed for bankruptcy as it lost its monopoly market position.

In 1987, JNR was split up and privatized, and it was relaunched as a group of private corporations after eighty years.

1.5 Discussions About Converting the Gauges Start up Again

Once the railways had been nationalized, the arguments about converting the gauges flared up again. An American railway financier named E.H. Harriman, who had visited Japan the year before the 1906 decision to nationalize the railways, strongly recommended making the railways standard gauge and even said that he was ready to provide 200 million yen of financing.

Harriman was one of the 'railway kings' who had accumulated great wealth for himself by building railways, and he was president of the Union Pacific Railway, which spanned the American continent from coast to coast. Harriman had extended his reach into East Asia and aimed to buy the South Manchuria Railway. Sensing the dangers in Harriman's offer, the Japanese government naturally did not respond to it.

At about the same time, plans were conceived to build a private railway in Japan, too, a high-speed railway that would link Tokyo and Osaka in six hours with EMU trains running on standard gauge tracks. The originator of the idea was one of the major financiers of the day, Zenjiro Yasuda, and he had actually brought together specific plans for construction and applied for a license to establish the 'Japan Electric Railway Corporation.' However, the government was in no position to acknowledge his application because it was also engaged in studies of the feasibility of building a standard gauge line between Tokyo and Osaka.

The first president of the Railway Authority, Shinpei Goto, was also enthusiastic about the idea of converting the track gauge. Goto, who had started out in the Superintendent-General's Office of the Taiwan Civil Administration and moved on to become the first president of the South Manchuria Railway, insisted that it was essential for the principal trunk lines, especially the Tokaido and the Sanyo, to be rebuilt with standard gauge tracks. He thought that this was needed for Japan's future economic development and for facilitating a unified transport system in which the South Manchuria Railway was united with the Japanese railways by way of the Korean peninsula.

In order to study this question more thoroughly, the government set up the 'Committee to Prepare for the Rebuilding of Broad Gauge Railways' in 1911. This committee compared standard gauge and meter gauge railways studied the economic effects of conversion to standard gauge and looked at the ideal future form of the railway network.

The committee was chaired by the Prime Minister himself, Taro Katsura, and Shinpei Goto, who was simultaneously president of the Railway Authority and Minister of Communications.

On a personal note, I'd like to mention that my wife's grandfather, an engineer for the Railway Authority named Fujita Tanaka, was a member of that committee. As the first director of the Operations Bureau, he had the highest responsibility for the departments that dealt with running the trains. He later became the fifth general manager of the Railway Technical Research Institute, but according to the memoirs of one of his subordinates at the time, he was very short-tempered and intimidating.

I myself was responsible for the Operations Bureau just before the privatization of JNR. I feel that there's some sort of destiny reflected in the fact that members of the same family were the first and last officials responsible for the operations of the national railways.

Yasujiro Shima, who made a huge contribution to Japan's rolling stock technology, was the superintendent, the top-ranking member of the railway engineers, and he, too, argued enthusiastically for conversion to standard gauge. From then on, the advocates of conversion were continually caught up in waves of politics. The Constitutional

Democratic Party agreed with the idea of conversion, while the Constitutional Party of Political Friends, including Prime Minister Takashi Hara, was critical of the idea. They advocated placing a priority on constructing new meter gauge lines rather than investing a huge amount of funds in converting existing meter gauge tracks to standard gauge. This was called the "Build first, convert later" argument, and it placed importance on the benefits to outlying regions.

Each time the administration changed, the arguments for conversion either disappeared from the scene or reappeared as if by magic, but the plans themselves never assumed a concrete form.

During that time, the corps of engineers laid down some standard gauge rails on the Yokohama Line, actually remodeled a locomotive to fit them, and conducted comparative tests on the performance of meter gauge and standard gauge locomotives. The experiment was probably intended as a demonstration for conversion, and of course, the standard gauge locomotive was more powerful and gave a superior performance, but in 1919, as the country faced the recession that followed the end of World War I, the Hara Cabinet called a halt to further discussions of gauge conversion, and for a time, this issue disappeared from the agenda.

From then on, the government followed the "build first, convert later" policy and promoted the construction of new lines. By 1930, the government railways had routes totaling 20,000 km, and construction of the main routes was nearly complete. It then became more important to improve main trunk lines such as the Tokaido and Sanyo and strengthen their transport capacity than to construct new lines into regions where there didn't seem to be any prospects of demand for transportation.

At this point, arguments about track conversion, which were supposed to have been over and done with, flared up again.

In 1940, the Railway Conference presented the government with plans to construct a standard gauge line between Tokyo and Shimonoseki, which is on the extreme western end of the island of Honshu. They received approval from the Diet and actual construction work began. This was the famous 'Bullet train plan,' a grand plan in which high-speed trains hauled by steam locomotives at 200 km/h would cover the distance between Tokyo and Osaka in four-and-a-half hours and between Tokyo and Shimonoseki in nine hours and fifty minutes, with future prospects of continuing on to the Korean peninsula via an undersea tunnel.

Work actually began on such projects with the Shin-Tanna Tunnel, the Nihonzaka Tunnel, and the Shin-Higashiyama Tunnel, but World War II broke out shortly afterward, and construction was halted. The dream of building standard gauge railways thus flared up and was extinguished several times. It was the dearest wish of Japan's railway officials.

Plans for building the bullet train resurfaced immediately after World War II. A group composed mostly of Keita Goto, president of the Tokyu Corp. and former officials of the Ministry of Transport, wanted to build a high-speed electric railway between Tokyo and Fukuoka with private funds in order to revitalize the economy, but it was absolutely impossible to bring these plans to fruition in the confusion of the immediate postwar period.

1.6 Competing with the World for Speed

High-speed railways were the perpetual dream of railway officials throughout the world, not just in Japan. It would be no exaggeration to say that the history of railway technology has been a story of challenging the limits of speed and safety. Beginning in the middle of the nineteenth century, soon after railways were invented, many dramas played themselves out around the quest for speed. I'd like to describe some of the most famous episodes in the history of railways.

Arguments about converting gauges occurred not only in Japan, but also in Britain, the country that invented railways. However, the arguments were not about converting meter gauge to standard gauge. In fact, the situation was the opposite of that in Japan: there were arguments about converting broad gauge to standard gauge.

The Great Western Railway, built in the western part of England by the brilliant railway engineer Isambard Kingdom Brunel was constructed with 2,040 mm gauge far wider than the standard gauge (1,435 mm). This 2,040 mm gauge allowed Brunel to run more powerful, faster locomotives, but this kind of railway was feasible only because the western part of England has few hills and is relatively flat. However, the construction of standard gauge railways proceeded in other regions of Britain.

In order to prove the superiority of broad gauge railways, Brunel built the *Great Britain*, a high-speed steam locomotive with just one large traction wheel on each side.

Hauling an express train, this locomotive made the 85-mile trip between Didcot and London at an average speed of 108 km/h. And this was in 1846! To provide some perspective, when you think about the fact that the *Super Hitachi*, the fastest limited express running on meter gauge tracks in the JR East system, maintains the same average speed—108 km/h—between Tokyo's Ueno Station and the city of Mito, you can see just how fast the *Great Britain* was.

However, despite such efforts, the last of the super-broad gauge disappeared from Britain in 1892 and were replaced by standard gauge tracks, because the majority of Britain's rail lines were already standard gauge.

In the 1870s, Britain's main rail network was almost complete, and there were two main lines running between London and Scotland's principal city of Edinburgh, one along the west coast and the other along the east coast. It was in 1870 that fierce competition for speed arose between these two lines. It developed into an almost daily competition to see which of the two trains leaving London simultaneously would arrive in Edinburgh first. The fastest express train on the east coast line was able to run the 632 km distance in seven hours and twenty-seven minutes, eighteen minutes faster than the scheduled time, at an average speed of 83 km/h.

Another speed competition occurred along the same routes in 1895. This time, the race was not from London to Edinburgh, but to the city of Aberdeen, much farther north and 850 km from London. On this occasion, the west coast train took the speed record, running the route in eight hours and thirty-two minutes at an average speed of 102 km/h.

These kinds of reckless competitions occasionally led to accidents and were soon halted. As the twentieth century loomed ahead, the speed record for steam locomotives was approaching 100 mph (160 km/h). In 1904, a steam locomotive called the *City of Truro* hauling an express train called *Ocean Mails* on the Great Western Railway posted a speed of 164 km/h during one of its runs.

Even before this, in the United States, the New York Central Railway's steam locomotive Number 999 is said to have reached a speed of 181 km/h while hauling the *Empire State Express*, but doubts have been raised about the truth of that claim.

In the twentieth century, the New York Central's *Twentieth Century Limited Express* and the Pennsylvania Railway's leading express train, the *Pennsylvania Special*, competed on the premier New York—Chicago route on the basis of speed and splendid accommodations, and both of them made the run in 18 hours. This route was over 1,500 km long, and the average speed was as high as 86 km/h.

Chapter 2
Trying for 200 km/h

Railways crossed the 200 km/h barrier surprisingly early, with a German 'Siemens train' reaching a speed of 203 km/h in 1903.

In 1955, the French National Railways achieved a speed of 331 km/h with an electric locomotive-hauled train. It is undeniable that Japan's railway technology at that time was at a very low level compared to the world's most advanced.

In this chapter, the author introduces the situation at that time when Japan was working hard to acquire AC electrification technology, one of the key technologies that made the *Shinkansen* possible, including his own experience of driving a steam locomotive.

2.1 Germany Was the First to Break Through the 200 km/h Barrier

Trains first broke the 200 km/h barrier earlier than you might think, in 1903.

On October 6 of that year, an EMU train built by the Siemens Company ran between the Berlin suburbs of Marienfelde and Zossen on a military rail line at a speed of 203 km/h.

Seventeen days later, an EMU train built by AEG posted a speed of 210 km/h. This was only 22 years after Siemens started running the world's first electric railway in those same suburbs of Berlin.

The electricity used was a three-phase AC at 15,000 V. Using three-phase AC meant that three electrical wires were strung over the tracks and that three pantographs attached to the roof of the train served as collectors, so by today's standards, it was a very unusual system.

The German Railway (DB) also turned its attention to internal combustion engines as a new motive force to replace steam locomotives. In 1933, it began running a two-car Diesel Multiple Unit (DMU) limited express train, *der Fliegende Hamburger*

© Japan Railway Technical Service 2024
S. Yamanouchi, *If there were no Shinkansen*,
https://doi.org/10.1007/978-981-99-8890-7_2

over the 270 km section between Berlin and Hamburg in two hours and eighteen minutes. Its average speed was 125 km/h, and its maximum speed was 160 km/h.

By this time, speeds of 160 km/h were not simply one-time speed records but speeds achieved during daily operations.

DB also built unusual rolling stock. In 1931, it built the *Schienen-Zeppelin*, a car resembling an airplane that had a propeller at its rear end and ran it over tracks. It posted speeds of 230 km/h, but of course, it was never put into actual service.

France's *Aerotrain* was a similar experiment, an air-cushioned levitating train powered by propellers and jet engines, invented by the technician of exceptional talent, Jean Bertin. In 1974, it clocked a speed of 430 km/h on the experimental tracks at Gometz, in the suburbs of Paris.

When I was working in Paris, I had the opportunity to ride this experimental train on the experimental tracks at Orléans. It really felt like riding in an airplane, and I think the speed was 380 km/h.

It was impressive to watch as we overtook and passed an express train of the French National Railway (SNCF), but the experimental train was horribly noisy.

At one time, there was talk of building a line like this between Orly Airport and the center of Paris, but the talk faded away when Bertin died. In the end, it was futile to talk about land-based transportation with propellers and jet engines.

A new age was dawning in which steam locomotives, which had played such an important role in the industrial revolution, were being replaced by new sources of motive power such as electricity and internal combustion, but the steam engines mounted one last challenge, aiming for speeds of 200 km/h.

In 1936, DB's streamlined locomotive, 05002, posted a speed of 200 km/h between Berlin and Hamburg.

Next, two years later in Great Britain, the birthplace of the railway, the London and North Eastern Railway's locomotive 4468 *Mallard*, hauling a special experimental train on tracks north of London, posted a speed of 203 km/h. This is the highest speed for a steam locomotive ever recorded anywhere in the world.

2.2   The Speed of Trains in the 'Underdeveloped' Railway Nation of Japan

Unfortunately, Japanese railways set no speed records of this sort until the *Shinkansen* arrived on the scene. The only run that even came close was in the postwar year of 1954 when a C62 steam locomotive posted 129 km/h on the Kiso River section of the Tokaido Line (Fig. 2.1).

This was not specifically a speed test, merely a test to find out whether the steel bridge over the Kiso River was strong enough to withstand the stress of a heavy locomotive running over it at high speeds. Yet the result was the highest speed ever recorded for a Japanese steam locomotive.

Fig. 2.1 Japan's most powerful steam locomotive, the C62. *Photo* Provided by the Railway Museum

It was in 1924 that Japanese railways formally determined the highest speed for a steam locomotive. Until then, no such determination had been made, because there had been no satisfactory speedometer available. The first locomotive to be equipped with a speedometer was the C51 in 1923.

At the time I began working for JNR, the speedometers fitted to steam locomotives often broke down. For that reason, part of the technical qualifying test for becoming an engine driver involved covering up the speedometer and having the aspirant estimate the speed of the locomotive. Every once in a while, the examiner said, 'Speed observation.' The would-be engineer then answered based on a visual estimate. The greater the difference between his answer and the actual speed, the more points were taken off the score.

In 1924, the maximum speed was specified as 95 km/h in the 'Acquaintanceship for Train Operations.' The fact that the rules for operating trains were referred to as 'Acquaintanceship' shows just what a bureaucratic institution JNR was.

In 1963, JNR completely modernized its rules and renamed them 'Standards and Regulations for Train Operations.' After JNR was privatized, the title reverted to 'Acquaintanceship for Train Operations' in order to bring it in line with the other private railways.

This rule about maximum speed was revised in 1947 so that speeds of 110 km/h were permitted on so-called special first-class lines, that is, main trunk lines. However, for a while, no trains actually ran at this speed because the track structures were weak and in no condition to withstand high-speed running.

The premier limited express train of the prewar period, the '*Tsubame,*' had a maximum speed of 95 km/h, but it took eight hours to travel between Tokyo and Osaka, so its average speed was no more than 68 km/h, about the speed of today's express commuter trains. During the twenty-two years between the opening of the Tanna Tunnel in 1934 and the completion of the electrification of the Tokaido Line in 1956, it was impossible to exceed this speed. The first train that could actually run at 110 km/h was the *Kodama* limited express, which went into service in 1958.

It is true that with its underdeveloped national rail system, Japan learned a lot of technology from the West and used it as the basis for technology suited to Japanese conditions, but there is no denying that the level of this technology was still quite a bit behind the rest of the world. A meter gauge railway system with lots of curves was, of course, destined to lag behind, but the technological sophistication of other Japanese industries was certainly at a similar level.

However, the South Manchuria Railway, which Japan built and operated in China, had standard gauge tracks, and its special limited express train, the *Asia,* which went into service in 1934, was able to travel the 701 km between Dalian and Xinjing (Changchun) in eight hours and twenty minutes at an average speed of 83 km/h and a maximum speed of 130 km/h. This was not the performance of the highest level, but it was nothing to be ashamed of when compared to railways in the West.

High-speed running requires a powerful locomotive and sturdy tracks, and that is why the meter gauge was at an absolute disadvantage. The output of a steam locomotive is determined by the amount of steam it generates. It needs a large boiler with a spacious firebox, and there are limits to how large an engine can be if it is to run on meter gauge tracks.

Wide tracks also have larger ties to support the rails, allowing them in turn to support heavier weights and providing better shock resistance during high-speed running. It is no wonder that Shinpei Goto, who had managed the South Manchuria Railway, and the well-informed technician Yasujiro Shima, placed their hopes for the future of the railway on widening the gauge of tracks.

High-speed tests began in Japan around 1955. At the end of that year, an EH10 locomotive and Naha 10 lightweight passenger coaches were used to conduct a speed test on the Tokaido Line between Kanaya and Iwata, and the resulting speed was 124 km/h. It was not a particularly impressive speed record, but it was significant as the result of what we might call the first thorough speed test.

In identifying the problems that are preventing increased speeds, the data that need to be monitored, and the mechanical equipment needed for monitoring, there are often many things that one cannot know unless one has devised and thoroughly tested. I can't rule out the possibility that this test was the beginning of preparations to build the *Shinkansen,* because both the EH10 locomotive and the Naha 10 passenger coaches were newly designed types of rolling stock completely different from the types then in service.

During that time, I was still a university student, but Hiroatsu Itoh, later chairman of JR Shikoku, who was scheduled to take a job with JNR at the same time as I was, invited me to come with him to see the tests.

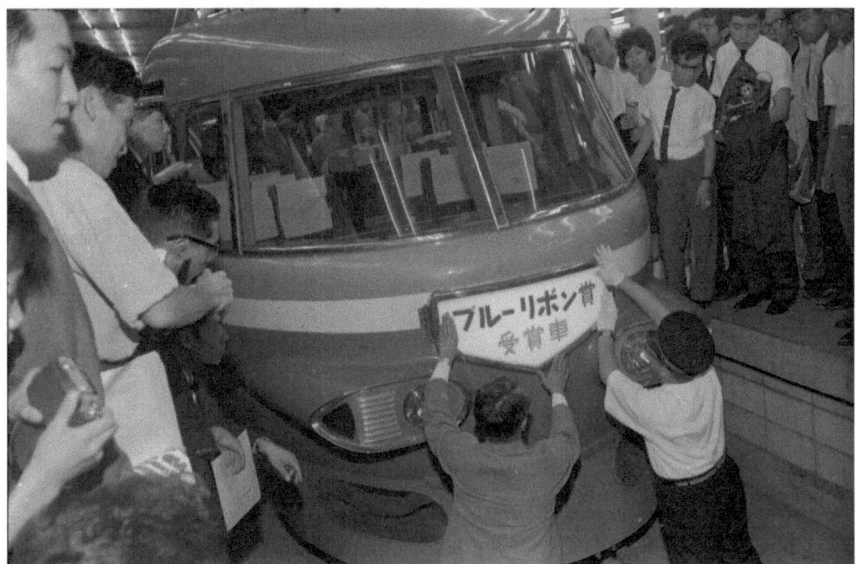

Fig. 2.2 Odakyu *Romance Car* NSE (Subsequent model to the SE) won the Blue Ribbon Award as voted by railway fans nationwide. (July 11, 1964, Odakyu-Shinjuku Station). *Photo* Provided by Kotsu Shinbunsha

The speed of the test train didn't leave such a great impression on me, but I was attracted by the earnest expressions of the engineers observing the test and by the way they talked about technology.

Among the technicians who were present were Dr. Tadashi Matsudaira, later managing director of the Railway Technical Research Institute, who was the leading expert on the problem of vibration in rolling stock, as well as Kyozo Kondo, Akira Hoshi, and Jyuzo Unoki. All of them played central roles in the design of the new rolling stock, especially the *Shinkansen.*

In September 1957, a high-speed test was conducted on the section of the Tokaido Line between Ofuna and Hiratsuka, this time with a new limited express EMU from the Odakyu Electric Railway called the SE type (Fig. 2.2). It achieved a speed of 145 km/h. The SE type was the newest type of EMU for high-speed use, and its articulation method was similar to that of France's later TGV, so it was the Japanese EMU train most suited for high-speed tests at the time. During this time, I had the opportunity to ride one of these experimental trains, and since the trip took place at night, I was impressed by how fast the lights seemed to rush past me.

After that, in 1959, JNR ran speed tests using the *Kodama*-type EMU, which posted speeds up to 163 km/h. In the following year, a special experimental EMU train ran at 175 km/h, and which was the Japanese record for meter gauge tracks (Fig. 2.3).

Fig. 2.3 EMU limited express *Kodama*, Series 151, which reached speeds of 163 km/h. *Photo* Provided by the Railway Museum

2.3   Impact of 331 km/h

An incredible record was set in 1955, the year before I began working for JNR. An electric locomotive of the French National Railways (SNCF) hauling three passenger coaches achieved a speed of 331 km/h. The location was a section of straight track south of the wine-producing city of Bordeaux, and CC 7107 was the first locomotive to post this speed on March 28. The following day, BB 9004 reached the same speed.

A French journalist friend of mine told me that the speeds were not actually identical and that CC 7107 was five km/h slower, but in order to preserve the manufacturer's reputation, the speeds were recorded as being the same.

My friend also showed me a photograph of the track that he had taken immediately after the test, and the rails were literally warped, an indication of the immense pressure exerted on them by the trains running at such high speeds. It was lucky that the locomotives had not derailed, and according to my friend, the engineer who drove the trains during the test turned pale when he saw the photograph.

However, these high-speed tests definitely influenced plans for the *Shinkansen.*

Those of us who had just joined JNR spent nearly a year at a dormitory educational facility called the 'Training Institute' and received new employees training. As part of our training, we were shown a film of the French high-speed tests. The way the pantograph of the locomotive continuously gave off huge sparks as it dashed along the track demonstrated to us how awesome it was to challenge technological limits.

What was then JNR's new training facility is said to have grown out of President Sogo's enthusiasm. He had a management philosophy that went right to the heart:

developing JNR begins with developing human resources. Unfortunately, however, most of the speakers who came to the Training Institute gave lectures that were far from inspiring.

Rather, it was the on-site training that was fresh and interesting.

As part of our apprenticeship as express train conductors, we made several round trips between Tokyo and Osaka, during which I had drunks pick fights with me. We also spent brief periods experiencing various kinds of work at Shinjuku Station.

One of my jobs was to stand at the entrance gate punching tickets, but before half the day was over, my fingers were covered with blisters, and there were so many passengers that I couldn't keep up with checking all their tickets and commuter passes. Sometimes the number of tickets sold and the money collected by the end of the day didn't match up. I asked an experienced ticket collector what I should do, and he told me, "You don't have to look at every ticket. You can tell if a person's cheating by his face and attitude."

Another job was working in the parcels office, spending the whole night sorting a huge number of parcels and comparing the documentation with the actual parcels. Each parcel had a number on it, and we were supposed to read that number aloud, but not in the usual way. For example, we read off parcel number 8241 not as *hachi-ni-yon-ichi*, but as *tako-futa-yon-choi*.

At Otsuka, I was put to work tamping down the ballast on the tracks with the pickaxes that were still in use then. The experienced workers went about their work gracefully, almost as if dancing, but an amateur like myself ended up hitting the sleepers or sending ballast flying. I was embarrassed as I sensed that the people on the platform were watching me.

2.4 Memories of Steam Locomotives

In our second year after joining JNR, we were split up for specialized training and began our real on-site apprenticeships. Those of us who had graduated from universities with degrees in mechanical engineering were assigned to work operating, inspecting, and repairing steam locomotives.

These days, people enjoy seeing mighty steam locomotives, and when one of them happens to be hauling a train, everyone gathers to watch it, but at that time, steam locomotives ran everywhere, and there was nothing unusual about them.

They were hardly ever seen at Tokyo Station, as you might expect, but the trains originating from Ueno Station on both the Tohoku and Joban Lines were entirely steam powered. Even at Shinjuku Station, small steam locomotives did the shunting work. I hoped that these kinds of things would soon disappear.

The trains hauled by steam locomotives tended to fill with soot and smoke, soiling people's faces and clothes, and besides that, they were slow. What bothered me even more was that I thought of them as symbols of the backwardness of JNR's technology.

Fig. 2.4 Working hard to shovel coal in the cab of a steam locomotive. *Photo* Provided by the Railway Museum

In my fourth year at university, I became an organizer for a students' educational tour of the manufacturing plants of major companies. Not one factory used steam engines. At Toyota, they told us, "We've succeeded in making the first real domestic cars" and showed me the initial models of the 'Crown' and 'Master.' Compared to the modern automobile assembly plant, the station yards and workshops of JNR looked like nothing more than relics of a bygone era.

My first job-training assignment was that of fireman. I was put in front of a mock-up of a steam locomotive to practice shoveling coal into the firebox; it looked easy, but it was really difficult (Fig. 2.4).

At first, I used a small shovel to toss in 100 or 200 buckets of coal, but by the second half of the training period, I had to toss in 400 buckets three times, a total of 1,200. Shoveling in more than 200 buckets made my wrists tired and stiff. Since the coal then fell off the shovel, I had to support the shovel by my waist and toss the coal in. If we dropped any coal, we lost points. We were also supposed to shovel in all the coal in a certain amount of time.

I had to do more than just shovel the coal into the firebox. The front half closest to the flapper of the firebox was called the 'back,' and it was important to pack a lot of coal in there. To do that, I had to turn the shovel over and slam it down toward me. I thought, "*This* is what I graduated from university for?" and that was how I honestly felt.

After two months, I was allowed to work on a real steam locomotive stoking coal. Maybe all that practice did me some good, because when I actually had to stoke coal, it went relatively well.

Fig. 2.5 Author driving a steam locomotive during his apprenticeship. *Photo* Provided by the Railway Museum

The locomotives that I worked on were the C59 and the D51 (Fig. 2.5). Unlike the practice period, we used large shovels, so I was able to shovel in a lot of coal at once. Once, just for the fun of it, I ran the engine continuously up an approximately 10-km gradient between Fukuma and Ebitsu on the Kagoshima main line with the safety valve operating on the whole time. This isn't something I'm proud of, but it just goes to show that I was no slouch at stoking coal.

I was also responsible for another bit of mischief.

One day, just when we were starting up, the conductor came to the locomotive and told us, "There's an executive from the Railway Administration Bureau today."

That's why I shoveled in a lot of coal and closed the blower just before we entered a tunnel.

This caused the passenger coaches to fill up with smoke.

In the third month, I was finally able to drive a locomotive.

Since I was assigned to Moji engine depot, that meant mostly driving express trains between Moji and Tosu on the Kagoshima main line. My job title was Apprentice Engineer, and a fully qualified engineer rode with me and taught me what I needed to know.

The man who served as my instructor was a wonderful fellow named Haruyoshi Yoshitsune. He was a real stickler when it came to work, but he was open and cheerful, and we often had good-natured arguments. If I got uppity and said something like, "When we think about the dynamics, it should be all right to go ten or twenty kilometers above the speed limit," he would scold me.

That's no good. You have to obey the rules.

On my first day driving a locomotive, I made the trip from Moji to Tosu without any trouble, and all of us, including the chief engineer and the assistant engineer, spent the night at the crews' lodgings.

It was a large tatami room with several futons laid out, and we all slept there in a row.

The next morning, Yoshitsune told me, "I had a terrible dream last night. When we go back today, I'll drive, not you." And so we left Tosu.

Just as we passed through Hakata and crossed the steel bridge over the Onga River, a human form suddenly appeared in front of the engine. We blew the whistle and applied the brakes, but we hit the person anyway. Even though I wasn't driving, it was a shocking thing to have happened on my first day.

The next day, I was driving, and we passed by a freight train going in the opposite direction on the steel bridge near Hakata. As the freight train passed us, I saw a line of elementary school children running across the track on the level crossing. I frantically blew the whistle and made an emergency stop. Fortunately, the children were all able to scatter to both sides of the track, and no one was hurt, but my hand seemed to be stuck gripping the brake handle.

Then on the following day, I had to make another sudden stop, this time near Futsukaichi, when I almost collided with a cart that was on the level crossing. After such a series of events, I was rather depressed, and that evening, I went for a walk with some friends in Mekari Park, which overlooked the Kanmon Straits between the islands of Kyushu and Honshu.

All of a sudden, I saw what looked to me like a man hanging from a large tree. Shocked, I ran back to the engine depot and told everyone, but an old engineer said to me, "Mr. Yamanouchi, you've just had a run of bad luck. Everything will be fine from now on."

For five months after that, nothing unusual happened.

After spending three months driving a locomotive, I could drive almost as well as a full-fledged engineer, as long as there were no unusual circumstances. Occasionally my instructor even said, "I'll leave everything up to you today. I'll be sleeping in one of the passenger coaches."

Although my skill was almost that of a full-fledged engineer, JNR didn't let me become one. I don't really know why, but there were rumors to the effect that if a 'bachelor' from a university caused an accident, it would look bad on his record, and they didn't want that to happen. I think it was an uncalled-for type of overprotection.

The use of 'bachelor' to refer to a university graduate was nearly obsolete in ordinary society, but JNR was an odd society in which the term was still alive. I think it was used as an insult meaning something like, "He lacks substance, but he's going to get promoted quickly anyway."

Even so, when my time driving locomotives came to an end, I took the actual skills tests. I was tested on operating time and the correctness of my stopping position, lack of jolting, and calling out signals correctly. When tested on a steam locomotive, I did well, scoring 88 points, but when I took the test for the electric locomotives on the Tokaido Line, I just barely passed with 60 points.

My times and stopping positions had been correct, and even if I do say so myself, I was able to drive the locomotive, so I thought the score was strange. When I asked the lead engineer who was proctoring the test to show me the scoring sheet, I saw that my score for 'attitude toward duties' was extremely bad.

A surprising thing happened when I took the skills test for the steam locomotive. I was about to start out as the first one to take the test when Yasuo Takano, Director of the Train Operations Division of the Western Directorate General of JNR, suddenly boarded the locomotive and said, "Step aside. This is going to be a model run."

This man was one of the 'bachelors' I looked up to. He drove just one section from Moji to Kokura and then handed the controls over to me, but his skills were wonderful to see.

Those were the kinds of capable professionals who worked for JNR in those days.

He was a zealous advocate of speeding up the trains, and I will say more about him later.

I studied steam locomotive inspection at the Otaru Chikko engine depot in Hokkaido. The famous manager Matsujiro Shimizu was there, and his workplace rules were very strict. On his own initiative, the employee assigned to do the inspection had to show up at the engine depot an hour or more before his scheduled work time and finish all preliminary inspections before the morning roll call. If we failed to do something, he would scold us in his Hokkaido dialect:"*Honde nee* (Well, now.)" We were intimidated by him and referred to him as '*Honde nee Matsu.*'

At the end of every work day, Shimizu had us gather around for a debriefing session.

Frankly, inspecting and repairing locomotives was not a pleasant job. I got dirty, and if I lost my concentration, I could break something. Since I was determined "to do the work of a full-fledged inspector," even though I supposedly lacked substance, it was tough on me. Both Shimizu and my colleagues were undoubtedly nervous about it.

The people at the work site were warm-hearted and friendly. Among them was Jiro Yamashita, who played a major role in reviving locomotive C62 3 later.

What annoyed me most was the drinking. No matter what happened, I'd hear, "Work ran late today, so let's have a drink," or "We finished early today, so let's have a drink," or "Strangely enough, we finished right on time today, so let's have a drink." Since everyone split the tab, I, as a non-drinker, was often asked to chip in. My task each day was to figure out how to slip away early.

At just that time, barbecuing first became popular in Sapporo. We had no sooner talked about it at the engine depot than we rigged up our own barbecue grill by making slits in a steel sheet, brought in a large bucket of mutton, and held a barbecue party right there in the engine depot. Those were the good old days.

2.5 Alternating Current (AC) Electrification

It was in 1956 that JNR began experimenting with AC electrification. The Senzan Line between Sendai and Yamagata was chosen for the experiment, four types of prototype locomotives were built, and finally, experiments began to see whether Japanese technology could claim this motive force for itself.

Those of us who had just joined the company were taken on a sort of field trip for new employees to visit the Senzan Line. I didn't understand any of the technology yet, but I could see that everyone was going through a lot of trouble to make the experiment work.

JNR set out to electrify its railways right after the war, beginning with the Tokaido Line and the Joetsu Line in 1946. By the following year, 1947, the electrification of the Joetsu Line had been completed.

One has to wonder at the desire and foresight to electrify the system when the rolling stock and the tracks were in a terrible state due to war-related damage and the harsh conditions of use during the war. If electrification work had not begun at this time, the electrification of the Tokaido Line would have been delayed for a long time, the development of electrical technology on Japanese railways would also have been slowed, and undoubtedly, the Tokaido *Shinkansen* would never have been built. In other words, this one bit of decision-making led to the completion of the world's first high-speed train.

Before the war, the only main trunk lines that were electrified, aside from some lines in the inner suburbs of Tokyo and Osaka, were the section of Tokaido Line between Tokyo and Numazu and the Kanmon Tunnel. Strong opposition from the military was the reason that electrification had not progressed any further. Their reason was that electrified rail lines are more susceptible to bombing.

When the war ended, that wall of opposition crumbled, and JNR set about electrifying the system. Another reason behind this decision was the fact that Japan was suffering from an extreme shortage of energy. There was no doubting the enthusiasm of the corps of engineers for electrifying the railways.

Electricity was far superior to steam as a motive force. In contrast to steam engines, which can use only five% of the energy from the coal they burn as the motive force, the electric operation uses 30% of their energy.

Yet electrical operations required a huge investment in facilities such as overhead contact wires and transformers. Therefore, a line could not make a profit unless there were more than a certain number of trains running on it. According to an estimate made at that time, one could make a profit on an electrified line if it ran 90 trains per day. At present,[1] 60% of JR lines are electrified.

The electrical power that Japan's railways was using at that time was 1,500 V of direct current (DC), but in the 1950s, Japan received the news that France had succeeded in electrifying railways with 50 Hz AC for general industrial use.

I'm going to get a bit technical here, but I'd like to explain the advantages of AC.

---

[1] As of 2019, the electrification rate of the JR Group was 62.5%. The slight increase was due to the opening of new Shinkansen lines.

The type of electricity that the electric company provides for us is called three-phase AC, and the frequency is 60 Hz in the United States and 50 Hz in Europe. Japan has both types, 50 Hz in the east and 60 Hz in the west. Furthermore, it is difficult to use this kind of electricity as a motive force without some modifications.

France and other southern European countries had electrified their railways with DC, while Germany, Switzerland, and other central European countries had used low-frequency AC. Like France, Japan was using 1,500 V DC.

Using DC, it is easy to make an electric motor with performance appropriate for driving electric locomotives and EMU trains. Yet, turning AC into DC requires setting up many transformers along the lines. And that's not all: once you decide on DC, you are no longer able to step-down the voltage with a potential transformer, so the transformer substations have to output a low voltage that the electric motors for the locomotives and EMUs can withstand.

The lower the voltage, the larger the current needed to yield the same amount of energy, and the losses of power along the way are greater. In addition, during high-speed operation, which requires lots of energy, the pantograph may be burned off, as happened during the French experiments.

On the other hand, with AC, a transformer built into the locomotive can step-down the voltage if high-voltage electricity is supplied, so this solves this problem. It had long been the dream of railway technicians to use electricity with the same frequency (called commercial frequency) as that commonly used in households and industry.

Hungary and Germany were the first countries to try their hands at this task. Beginning around 1915, Ganz, the powerful Hungarian manufacturer of heavy electrical equipment, produced ten locomotives for 50 Hz, single-phase AC and repeatedly tested. In 1931, the 190 km line running from the Hungarian capital of Budapest to the Austrian border was electrified using 15 kV 50 Hz commercial frequency, thus succeeding in the world's first practical application of the technology.

In 1936, Germany electrified the 56 km Herrental Line, a mountainous line in the southern Black Forest with 20 kV 50 Hz commercial frequency and built and began tests on four different types of locomotives. After World War II, France occupied this region, and the corps of engineers from SNCF, who visited the line, became interested in the technology. France, too, then started testing electrification with commercial frequency AC, choosing the Annecy Line in the Alps as its test line.

France played a principal role in technological innovation for railways after the war. Having lost the war, its erstwhile rival, Germany, was preoccupied with recovering from war damage and was in no shape to pursue technological innovations. Rather than modernizing its railways, Britain was moving more in the direction of shrinking and rationalizing its rail network.

Just as the *Shinkansen* came into being under the strong leadership of President Sogo and Chief Engineer Shima, France's ventures into new technology occurred under the leadership of the then-President of SNCF, Louis Armand.

He was active in the French resistance against Germany during the war and was arrested on June 24, 1944. The film *The Fight for the Railway* is said to have been modeled after him. He was a young forty years of age when the war ended in 1945, and he became the general manager of the office in charge of rolling stock. Four years later, he assumed the post of president of SNCF.

2.6   Elite Education in France

France's elite education is famous, and it can be said that along with graduates of the École Nationale d'administration (ENA), the gateway to success for high-level government officials, graduates of the Ecole Polytechnique rule over the worlds of government and business.

The Ecole Polytechnique is a school founded in 1794 for the purpose of 'providing the scientific and general education necessary for shouldering high rank and heavy responsibility in the scientific, technological, and economic areas of the government, the military, and the business sector.' About 400 students enter every year. About one-fifth of those who take the entrance exams are accepted, but the difficulty of the exam and the difficulty of the curriculum after admission cannot be compared to that of the University of Tokyo.

Instead, graduates of this school literally become a privileged class. They are called '*X's*,' and when I was living in Paris around 1970, all except one of the general managers of the offices within the headquarters of SNCF were *X's*.

The best students among the *X's* are called *bottiers*, and they are afforded a different class of favorable treatment. Armand was a *bottier*, and after graduation, he did geological research in the military. When at age 29, he took a job with the National Railway, his first position was already that of department manager.

After becoming President, Armand was most interested in high-speed running and electrification of the railways. It was under his leadership that the speed record of 331 km/h was set.

With the experiments on the Annecy Line, SNCF had set the goal of practical application of electrification with AC, and serious work began on electrifying the main line in northern France, which carried a lot of freight traffic.

At about this time, JNR also became interested in this technology. JNR President Sounosuke Nagasaki visited France in 1953, and upon returning to Japan after talks with President Armand, he set up the "Survey Committee on Electrification with AC," which was the beginning of serious study of the matter.

Yet as I explained before, AC electrification at commercial frequency was quite difficult from a technological point of view. I found the following information in some internal documents from that period:

> Since it would probably be difficult to produce AC locomotives in Japan right away, we decided to conduct the tests on the Senzan Line with imported locomotives, and to build only prototypes of the major devices for establishing a future manufacturing base. We also decided to attach domestic prototypes of an AC electric motor and mercury rectifier to a scrapped DC electric locomotive and at the same time, to make prototypes of transformer control devices, etc., and set up two test trains.

As far as imports are concerned, we will import two examples of a single model of AC motor locomotive along with the manufacturing rights for them, because they are supposed to be the items most difficult to manufacture in Japan. As far as the mercury rectifier locomotive is concerned, we have acquired some confidence in domestically manufactured mercury rectifiers for rolling stock, so we will take care of this with a domestic prototype.

The documents continue, "discussing what would happen if there were no prospects of importing AC locomotives."

As long as we cannot confirm our ability to order 10 to 15 locomotives in succession, we think that we have almost no prospects of importing a sample locomotive. Therefore, we will write to SNCF and ask them without conditions if they will be so kind as to let us purchase one AC locomotive and one mercury rectifier locomotive for the purposes of testing.

In May of 1955, SNCF used the occasion of completing its first real AC electrification section between Valenciennes and Thionville (271 km) to sponsor an international conference on AC electrification in the city of Lille.

This region close to the Belgian border is France's prime industrial zone, and in fact, Toyota has built a factory near Valenciennes.

Five representatives of JNR participated in this conference, and after it ended, they entered into negotiations for importing an AC locomotive, but as supposed, they were unable to come to an agreement.

The French manufacturers are alleged to have said, "If we sell them one or two locomotives, all they'll do is imitate our technology. We won't sell anything less than a significant number of locomotives."

What else could they have expected? The result was that Japan made all the prototype locomotives on its own.

There are many types of AC locomotives. From a technological point of view, the type in which the AC is not converted to DC but moves the locomotive directly with an AC motor is attractive, but the technology is quite difficult.

For the experiments on the Senzan Line, four types of locomotives were built and tested: two locomotives in which the AC as changed to DC with a mercury rectifier, and types in which the current moved an AC motor directly. As a result, JNR concluded that the mercury rectifier mode was best for Japan.

Thus locomotives with mercury rectifiers were adopted for Japan's first AC trains, which began service between Maibara and Tsuruga in 1957.

The advantages of electrifying rail lines with AC are lower construction costs, great traction power because the locomotives don't slip when starting up, and the ability of even a small locomotive to haul heavy freight trains, so it is not necessarily true that this technology was developed only for high-speed running. However, since using AC makes it easy for the pantograph to collect current, this was an essential technology for the *Shinkansen*.

Initially, it was determined that the *Shinkansen* would use 60 Hz AC, but many technological questions remained. Would it be hauled by a locomotive? Would it be an all-EMU train? If so, should JNR try to develop a system using an AC motor directly? Or should it use a mercury rectifier system? Mercury rectifiers have huge

Fig. 2.6 An AC/DC commuter EMU train on the Joban Line today. The insulators on the roof are very impressive. *Photo* Provided by Testuro Aikawa

dimensions, and they were all right for locomotives, but they were difficult to use with all-EMU trains.

At just that time, a rectifier employing a silicon diode was invented, and the first AC/DC EMU train using this technology began running on the Joban Line in 1960 (Fig. 2.6).

With this technology available, it became possible to envision the *Shinkansen* as an all-EMU train with a silicon rectifier.

Looking back, we can see that the new technology became practical just in time for the *Shinkansen* to become a reality, and we can also find out how many important decisions and human efforts lay behind this venture.

Chapter 3
The Curtain Rises on the Era of Express Trains

Japan's high-speed trains are all EMUs. This is quite exceptional on a global scale. Locomotives usually haul long-distance trains in many countries in Europe and Asia, and high-speed trains such as TGV and ICE, the Japanese equivalent of the *Shinkansen*, were no exception.

One of the reasons for the development of EMUs in Japan was partly that the number of arrival and departure tracks at major terminal stations in Japan was small, making it more advantageous for EMUs that did not require turnaround tracks.

In this chapter, the author will introduce these circumstances: the majority of JNR's main trains are the power-distributed EMU and DMU. The story around it is introduced with various episodes.

In this period, the author, as a young officer, was the head of the operations depot where a large number of DMU limited express trainsets were put into service for the first time, and while confronting the labor union, he describes the modernization of JNR at that time through his contact with staff at the worksites.

3.1 Japan, the Kingdom of Electric Multiple Unit (EMU) Trains

Japan's *Shinkansen* is made up entirely of EMUs. That's not all: almost all the limited express trains on the non-*Shinkansen* lines are made up of EMUs, and just about the only passenger trains hauled by locomotives are sleepers on overnight runs or chartered trains for groups with cars furnished in traditional Japanese style. If we look at railway systems around the world, we find that this is an exceptional situation—one might say an almost unique situation.

In most European and Asian countries, long-distance trains are usually made up of passenger cars hauled by locomotives, and even France's TGV and Germany's ICE, the high-speed trains corresponding to Japan's *Shinkansen*, are no exception. They

© Japan Railway Technical Service 2024
S. Yamanouchi, *If there were no Shinkansen*,
https://doi.org/10.1007/978-981-99-8890-7_3

may look like EMUs from the outside, but they are actually effectively composed of a string of passenger cars hauled by a power car.[1]

In China, there are hardly any EMU trains at all if we disregard the subways in such cities as Beijing and Shanghai. Trains for intercity travel are all made up of passenger cars hauled by locomotives.[2]

Japan's privatized JR companies and subways together own 46,000 EMU cars, which makes it a 'kingdom of EMU trains' like nowhere else in the world. But how did it get to be this way?

When I joined JNR in 1956, not only the limited express but also almost all superior trains, such as express trains and semi-express trains, were made up of passenger cars hauled by locomotives. EMU trains ran only on the urban transportation systems for cities such as Tokyo and Osaka or on private regional railways. The only exception was the Shonan EMU train which went from Tokyo to the Izu Peninsula. Its range exceeded 100 km, and semi-express trains such as the *Ideyu* were also eventually added to the fleet. It would be fair to say that the Shonan EMU trains were pioneers in turning Japan into a 'paradise for EMU trains.'

Until then, EMU and DMU equipped heavy electric motors or internal combustion engines, which caused severe vibration and an uncomfortable ride. Add to that, the loud noise they produced, and it's no wonder that the conventional wisdom in the railway world was that these kinds of trains could not be used for long-distance trains.

Sometimes social changes and technological advances make common knowledge seem not so wise. The notion that EMU trains couldn't be used for superior trains with fare supplements was one such bit of conventional wisdom.

Some people say that the Shonan EMU train grew out of advances in EMU technology, but that isn't necessarily the case. From a technological point of view, the Shonan EMU trains were still old-style EMU trains. When I asked Akira Hoshi, who was active as a train designer in those days, he said, "Certainly, they were what you might call old-style trains, judging by their technology, but the way that incorporation of passenger car planning concepts into the chassis' layout was a real milestone. It had the entrance and exit separated from the passenger compartment, and it was also nicely equipped with a toilet."

You might say that EMU trains thus took their place as full-fledged members of the long-distance passenger fleet. Until then, EMU trains were considered one step below locomotives technologically, and they were also considered to be one rank below unpowered passenger cars in terms of their interior facilities (Fig. 3.1).

I once worked aboard the *Ideyu* on the Shonan EMU train as an apprentice conductor, and it vibrated quite a bit. However, the vibration wasn't bad enough to make the train unworthy of designation for the superior train. At that time, even passenger cars didn't provide a smooth ride, and sometimes the waitresses trying to

---

[1] In recent years, France has been introducing EMUs and DMUs with three to seven cars in the Transilien and Transport Express Regional (TER), which connect Paris to the suburbs and dual-mode MUs (AGC), which operate directly on electrified and non-electrified lines.

[2] In China today, high-speed rail is operated by EMUs.

Fig. 3.1 Shonan EMU train, which revolutionized EMU trains. *Photo* provided by the Railway Museum

balance trays of food in the old-style dining cars looked as if they were performing in a circus.

It might be said that the Shonan EMU train grew out of uniquely Japanese conditions. Yonehiko Ishihara, General Manager of JNR Train Operations Bureau in the early 1960s, and an ardent proponent of high-speed running, reminisced about that era.

The ones who designed the Shonan EMU train and put it into service were Torazo Kijima, General Manager of the Transport Bureau, and Hideo Shima, General Manager of the Rolling Stock Bureau. At that time, there was this idea that 'EMU trains can run up to 100 km and be made up of 15 cars. What do you mean? Those kinds of things simply have to be done with locomotives.' It was what you might call the common knowledge. I felt as if they were saying, 'Are there any trains like that anywhere in the world? What are you saying? Are you talking in your sleep or something?'. At that time, Shima got the idea of distributed power, which meant having only EMU trains run on the entire Tokaido Line in the future. What finally decided the solution was that Kijima said, 'The way things are going, it will be impossible to transport anything on the Tokaido, you know. If we have to do it with locomotives, no matter what, we'll have to have another platform at Tokyo Station and another locomotive turn around track. Then Shima said, 'We can't do that.' And then Kijima said, 'If we can't do that, then we have no alternative but to run EMU trains.' And that's why they decided on EMU trains.

Beginning in 1953, EMU train technology entered a new era of innovation. Technology usually just moves forward in a steady but unglamorous manner, but every once in a while, it makes a great leap.

3.2   New High-Performance EMU Trains

The privately owned railway companies got into technological innovation earlier than JNR. First, the weight of the carbody was reduced by 30%. That was because they were able to make the outer steel plates of the carbody thinner through the use of high-tensile steel and lightweight alloys, and at the same time, they adopted a design method called a monocoque structure, which strengthens the whole structure of the carbody.

In the old days, passenger coaches were all made of wood, so the strength of the carbody was supported by an underframe comparable to an automobile chassis. That concept didn't change, even after manufacturers began to make the carbody out of steel, so the cars became extremely heavy.

In contrast, in the body of an airplane, it's the monocoque structure that gives strength not just to the frame but also to the body itself, but adopting that structure for an EMU, which has large windows and doors, was rather difficult.

Many of you may already know that one of the first EMU trains to employ this kind of new, lightweight car body was the green EMU, nicknamed the 'Green Frog,' that used to run on the Tokyu Electric Railway Company.

The first to take up the challenge of trying to use this structure was the French Northern Railway Company, but the first country where it seriously adopted was Switzerland. In a country like Switzerland, with its steep mountains and lots of curves, the effect of reducing the weight of the passenger cars was strikingly large. Swiss Federal Railways (SBB) and the Schlieren company created a special research team, and after amassing five years of studies, they began running the new, lightweight passenger cars on express trains between Zürich and Geneva. After World War II, lightweight cars became the standard type for SBB.

In 1953, JNR engineer, Akira Hoshi, spent a year in Switzerland and Germany studying lightweight railcars. On its own, the Railway Technical Research Institute also developed a new method for calculating the strength of a car body. The result was the Naha 10, JNR's first lightweight passenger car, which was used in high-speed tests on the Tokaido Line in 1955. In contrast to previous passenger cars, which weighed 34 tons, the new ones were lighter at 28 tons.

The power device for the EMU trains also changed significantly.

Previous EMU trains had a large electric motor mounted directly over the wheel axle, and in order to minimize the shock on such a large and important component, it was best to mount it on a spring on the bogie. That also minimized the shock applied to the rails during high-speed running. However, when you mount an electric motor on a spring, you have to worry about the gears not meshing properly due to vibration. The electric motor on an EMU revolves at high speed, and it turns the wheel axle by means of the gears.

In contrast, around the beginning of the twentieth century, a technology was developed in countries like France and Switzerland in which a flexible coupling similar to the propulsion shaft on an automobile was used to transmit the torque from the electric motor to the wheel axle. In that case, it doesn't matter if the distance between the electric motor and the wheel axle changes somewhat, so you can put it on top of a spring. It seems insignificant, but it was a very important technological advance.

Furthermore, there were developments in the electric motors themselves, including a high-speed electric motor capable of 1,500 revolutions per minute (rpm). Up to then, the speed had been about 900 rpm. When the electric motors became faster, you could make a smaller, lighter motor with the same output.

In addition, until then, each car of the EMU train had all the drive apparatus, but around this time, the design changed so that two cars were linked as one unit. That let designers reduce the number of control mechanisms and other components. The brakes were changed to electromagnetic straight brakes, which were much easier to operate than the old-style ones, and worked much better.

The greatest advance was the bogie structure of the car. The bogie is supported by double springs in order to soften the vibration of the car, but if you just leave it at that, tremendous vibration arises at a certain speed. That's the phenomenon called resonance. In order to prevent that, the old-style bogies used leaf springs to prevent this resonance. However, leaf springs tend to bounce, so they don't necessarily yield a satisfactorily smooth ride. The Shonan EMU trains still used this type of bogie.

The use of oil-filled dampers began at about this time, and advances were being made in analytical theories of vibration as applied to bogies, so it became possible to build bogies that provided an extremely smooth ride with only coil springs.

Another problem is the prevention of hunting, a type of vibration in which the car swings laterally like a moving snake. If the bogie is not well designed, hunting occurs above a certain speed. The principal cause of hunting lies in the slight clearance between the axle box that supports the wheel axle and the frame of the bogie (Figs. 3.2, 3.3).

Engineers in countries such as France, Germany, and Switzerland carried out research to see if they could build a bogie without this kind of gap, and the result was a variety of new kinds of bogies. In Japan, too, new types of bogies incorporating technology from other countries and new types of independently designed bogies that hunted very little appeared on the scene. A typical EMU train incorporating the new design methods of that era to the fullest extent is the red train called the 300 Type, which ran on the Marunouchi subway line in Tokyo until recently.[3] This EMU train had the electric motor mounted on a spring in the bogie, and the wheel axle was driven by a special mechanism using a gear called a WN drive. This type of apparatus is also used on the *Shinkansen*, but the 300 Type was the pioneering rolling stock employing this technology.

These types of EMUs, with their completely new design, are called 'new high-performance EMU trains.' The first new high-performance train on JNR, the *Moha* 90 Type (later the 101 Type), appeared in 1957. This was somewhat behind the private railways, and one can't help feeling that the design was a bit conservative as well, but once JNR decided on a standard model, it was manufactured in large quantities,

---

[3] This railcar was taken out of service in 1996 and was then utilized on urban railways in Buenos Aires, Argentina.

Fig. 3.2 A EMU train of 300-type cars on the Marunouchi Line of Eidan subway system (now Tokyo Metro), which incorporated new design features. *Photo* provided by the Railway Museum

Fig. 3.3 JNR's first new high-performance EMU train, the 101 type.
Photo provided by the Railway Museum

so in any case, the tendency was to be conservative. However, this 101 Type was a groundbreaking EMU that changed JNR's EMU technology significantly, and it is no exaggeration to say that this accumulation of technologies led to the *Shinkansen*. In all, 1,535 of the 101 Type were produced, but if one groups them together with the 103 Type, which came later, 5,000 of them were made.

3.3 The *Kodama* Limited Express Appears on the Scene

The limited express version of the new-efficiency EMU trains was the *Kodama* (the 151 Type EMU), which first appeared in the fall of 1958.

This EMU train went from Tokyo to Osaka in six hours and fifty minutes, or forty minutes faster than the *Tsubame* and *Hato* locomotive-hauled limited express passenger trains. It was nicknamed the 'business express' because you could do some work in Osaka—even if it was only for two hours—and make a roundtrip between Tokyo and Osaka in one day. And the faster running time not only made it more convenient but also allowed each EMU train to make one round trip per day, which led to more efficient use of rolling stock. It could travel a thousand kilometers per day. The increased speed also had a great effect on management.

The 151 Type was Japan's first real limited express, and it was also a breakthrough EMU train that allowed EMU trains, which had previously been considered inferior to locomotive-hauled passenger trains, to take a leading role. One could even say that it was this train that built the foundation of this kingdom of EMU trains, which is extremely unusual when compared with the rest of the world.

However, one could also reasonably say that from a technological point of view, it was nearly the same as the 101 Type except for the use of air springs. In that sense, one may be able to say that it was the 101 Type, or perhaps the noticeably earlier new trains on the private railways, that created this kingdom of EMU trains.

It was in Japan that air springs were first adopted in rolling stock, having been developed by Kisha Seizo, Ltd. (now absorbed into Kawasaki Heavy Industries, Ltd.). It is said that an executive from that company was on a business trip to the United States when he saw that the intercity Greyhound buses used air springs, and he thought about using them on rolling stock.

The private Keihan Electric Railway was the first to use air springs, employing them in a limited express EMU one year before the *Kodama* appeared.

Another feature of the *Kodama* that was the result of the great effort was its air conditioning. Up to that point, only some sleeping cars and dining cars had air conditioning, but it was the first train to have air conditioning in all cars.

Akira Hoshi recalls that era:

When Shima came back to JNR as chief engineer—it was around 1956. Anyway, window type air conditioners of consumer-model began to appear on the market, and after Shima proposed adapting them for use on trains, dispersal-type ceiling air conditioners were developed. Though a large centralized unit air conditioner was used in passenger cars, this mode of air conditioning was found only in Japan, and now it has become the standard for air conditioning in passenger trains throughout the country. It may be noted that the *Shinkansen*'s heat pump type of air conditioning, which can also be used as a heater, was the first of its kind in the world.

3.4 Why EMU Trains Developed

Japan had its own unique reasons for becoming the kingdom of EMU trains.

First, as I have already mentioned in discussing the Shonan EMU trains, Japan's railway terminals had insufficient facilities for the number of trains they served.

Tokyo has three terminals, Tokyo, Shinjuku, and Ueno, but in fact, nearly all the traffic concentrates at Tokyo Station. In contrast, cities such as London and Paris have several terminals.

Three thousand seven hundred trains go in and out of Tokyo Station every day, but there are only ten tracks. Until September 1997, the Tohoku and Joetsu *Shinkansen* ran 240 trains in and out of the station per day from only one platform. After the passengers get off a train arriving from Morioka, there is only a twelve-minute interval to clean the interior of the train before it once again heads north. Railway executives visiting Japan from overseas are uniformly amazed by this kind of precise technology for operating trains.

About 300 trains come and go every day at Lyon Station in Paris, but the station has 22 tracks. Madrid's Atocha Station, which is the terminal for the AVE, Spain's counterpart to the *Shinkansen*, has seven platforms, but there are only 68 trains per day.

In the old days, Japan, too, took a more leisurely approach. The former *Tsubame* limited express used to deadhead to Tokyo Station from the Shinagawa rolling stock yard 27 minutes early and wait for its departure time. After construction of the Tokaido *Shinkansen* began, the platform that the *Tsubame* had been departing from was unavailable, due to construction work, so it became a challenge to figure out how many minutes the *Tsubame* needed to prepare for departure, and I was put in charge of investigating the issue.

When I went to Tokyo Station, I found that taking the locomotive that had hauled the deadhead train into the station and hooking it up to the other end of the train was the job that took the longest. If we had another locomotive ready and waiting at the outbound end, it would be possible to board the passengers and depart in five minutes. Still, going from 27 to five minutes seemed rather harsh, so I went back with the message that shunting could be completed in 10 minutes. I remember that the director scolded me, "Hey, what do you know about limited expresses?".

That demonstrates the high status held by limited express trains in those days.

If that director could visit Tokyo Station today and see how *Shinkansen* trains that have deadheaded into the station are ready to leave in four minutes, I wonder what he would say?

This kind of dexterity is not possible with locomotive-hauled passenger trains, which require a locomotive to be replaced every time the train turns around. This is the first reason that Japan's EMU train mode of transportation developed.

However, you may be wondering about the practice of attaching a power car to both ends of the train, as is done with France's TGV and Germany's ICE. That wouldn't be impossible, but there are no traces of this method ever having been considered for the Tokaido *Shinkansen*.

JNR Chief Engineer Hideo Shima, who chaired the Survey Committee on Construction Standards for the *Shinkansen*, was a strong advocate of the EMU train configuration, and he propounded the idea that EMU trains were best even for freight use (Fig. 3.4).

On the other hand, some members of the committee were of the opinion that using electric locomotives for hauling freight trains was more economical, and in that case, that using electric locomotives to haul local trains (today's *Kodama* in *Shinkansen*) would make for more efficient use of these locomotives.

Shima gave the following reasons for promoting the all-EMU train configuration: it would be possible to lighten the axle weight (the load applied to one axle of the car), there would be no need for extra tracks to allow for attaching the locomotives, and it would be possible to use electric brakes.

All his life, Shima held a firm belief in the EMU configuration and in having all railcars in a trainset to be powered with electric motors. I myself also received a copy of 'The Argument of All Rolling Stock to be Powered with Electric Motors' directly from him. I can't help feeling that Shima, who in his younger days had played a

Fig. 3.4 Hideo Shima.
Photo provided by
Kotsu Shinbunsha

part in designing steam locomotives, including the D51, was keenly aware of the limitations of locomotive-hauled trains.

Certainly, in terms of performance, EMU trains make it easy to increase output, and the idea of being able to use electric brakes fully is attractive. However, as the number of powered devices increases, the price of the rolling stock goes up. The reason that Europeans don't seem very eager to use EMU trains, except in subways and other urban transit systems, is mainly the problem of cost.[4]

Europeans use a type of train in which the locomotives that haul the passenger cars are not detached and reattached at the other end but instead push the train from the other end. These are called 'push–pull' trains. The TGV and ICE, which have power cars at both ends, are not the original style of push–pull train. This type of train ran on the suburban lines around Paris beginning in the 1930s.

When a locomotive pushed a train from the rear end, the engineer operated the steam locomotive by remote control from a driver's cab in the coach at the opposite end of the train. I think it was a very original idea. This use of a broad, standard-gauge track, and a shock absorber, called a 'buffer,' on the end where the locomotive is attached allowed the train to run safely, even when pushed from the rear. Some of the push–pull express trains run at up to 160 km/h.

In Japan, the sleeper trains that arrive at Ueno Station are then pushed to the depot at Oku, but the speed is kept down to 45 km/h. Since there is no specific engineer's cab in the coach at the opposite end of the locomotive, it is equipped with a simple brake apparatus and a warning whistle.

3.5 The Era of Limited Express Trains

The era of limited express trains came to the conventional meter-gauge lines just before the completion of the *Shinkansen.* The high rate of economic growth during that period made these trains a necessity, and so did the prospect of competition with two growing means of transportation, automobiles and airplanes.

I think that the opening chapter of the postwar era of limited express trains may have taken place in 1956. It is common knowledge that Japan's first limited express trains were put into service in 1912 to run between Shinbashi and Shimonoseki, but these limited expresses disappeared during World War II, to be resurrected after the war in the form of the *Tsubame* and *Hato,* which ran between Tokyo and Osaka, and the *Kamome,* which ran between Kyoto and Hakata.

The year 1956 saw the launch of a new limited express sleeper train, the *Asakaze,* running between Tokyo and Hakata. It ran on the same section as the previous limited expresses, but it went against the former conventional wisdom by passing through Osaka in the midnight. It was Torazo Kijima, General Manager of the Transport Bureau, who came up with that idea. He is alleged to have said, "As Japan goes on

---

[4] In recent years, Europe has also been introducing EMU-based high-speed trains, such as Germany's ICE3 and ICE4 and Italy's Italo.

recovering, everyone will be linked through Tokyo." The result was that this new train was five hours faster than existing express trains.

There was a great deal of strong opposition to this new limited express sleeper train within JNR. The Business Bureau of the Headquarter was evidently dismissive: "A train that ignores Osaka simply won't be viable." The Osaka Railway Administration Bureau was also opposed: "If there's a train that passes through Osaka at midnight, we won't let it come through Osaka Station."

To counter this, the executives of the Train Operations Bureau at the Headquarter, who were in favor of this train, said, "If that's your attitude, we'll just send it around on the northern freight line," so the Osaka Railway Administration Bureau grudgingly accepted the situation.

The busiest major terminals, such as Tokyo and Osaka, have dedicated bypass lines for freight trains so that they end up not passing through the station. "The northern freight line" is the designation of the freight lines that avoid Osaka and pass through the Miyahara marshalling yard.

The new *Asakaze* was made up of a collection of old-style passenger cars, so it didn't look very elegant, but it was a huge success, and afterward, there were as many as seven limited express sleepers per day making the round trip between Tokyo and Kyushu.

The next step came in 1958 with the *Kodama* and *Hatsukari* limited expresses.

The *Kodama* was the breakthrough train that shattered the fixed idea that EMU trains weren't suitable for use as limited expresses due to their inability to provide a comfortable ride. The *Hatsukari* on the Tohoku Line was hauled by a steam locomotive for the whole section, but it was noteworthy for being the first limited express to appear somewhere other than on the Tokyo—Osaka—Kyushu route.

It marked the first time that a new limited express route had been added since 16 years earlier in 1942, when the Kanmon Tunnel was opened and the *Fuji* limited express was extended to Nagasaki. At the time that the *Hatsukari* was put into operation, there were arguments about whether a limited express could be viable on the Tohoku Line, but it went into service anyway, no doubt because the success of the *Asakaze* had raised everyone's enthusiasm. However, it is also true that the *Hatsukari* didn't have very many passengers for a while.

Another reason that the *Kodama* was such a breakthrough was that it was the first limited express in Japan that was able to travel at a maximum speed of more than 100 km/h.

As I mentioned before, the 1947 revision of the 'Acquaintanceship for Train Operations' contained a rule stating that the maximum speed on the main trunk lines known as 'special first-class lines' was 110 km/h, but actually, no train ran at that speed, because the tracks couldn't stand up to very high-speed operation.

The new 151 Type EMU train used for the *Kodama* was not only much lighter than that of a locomotive but the cars also had a bogie that didn't sway very much and therefore didn't apply much shock to the tracks. In order to make it possible for the *Kodama* to run at the 110 km/h speed that the rules allowed, JNR improved the tracks, contact wires, and signal facilities.

The purpose of the 1955 high-speed tests was to gather data for those improvements because not only the *Kodama*'s maximum speed but also its speed when passing through turnouts was enhanced five km/h faster than that of other trains. As a result, a time savings of 40 minutes was attained.

In those days, a common catch phrase was 'one minute, one hundred million yen.' That meant that reducing a train's traveling time by one minute required one hundred million yen of investment in facilities.

By the way, if you're wondering whether this catchphrase applies to the Tokaido *Shinkansen*, each minute of time shaved off its run required 1.9 billion yen worth of investment. Of course, the goal of building the Tokaido *Shinkansen* was actually to increase transport capacity more than reduce the traveling time.

Around this time, JNR began seriously dealing with not only the construction of the *Shinkansen* but also with modernizing the conventional meter-gauge lines. First, in 1958, it established the 'Motive Power Modernization Committee,' which investigated the question of how to proceed with replacing the 6,000 steam locomotives, which were still the main source of motive power, with electrical motive power operation and diesel locomotives.

This involved more than just changing the motive power source. The committee grappled with several challenging problems. They needed to know what kind of rolling stock to produce: locomotives, EMU trains, or DMU trains? How should rolling stock depots/yards be distributed? How should employees be trained and where should they be reshuffled? Finally, how much financial investment would be required, and what would be its effects?

In a sense, it was even a plan for restructuring the entire JNR. In the ensuing one year and five months, the committee met more than 40 times, and it issued its findings in 1959. It concluded that "Within 15 years at the latest, JNR should electrify its approximately 5,000 km of main trunk lines, convert the remaining sections to diesel, and get rid of all its steam locomotives. We recognize this as not merely a necessity for managing JNR, but also as an extremely desirable measure for the economic wellbeing of the Japanese people."

JNR's last steam locomotive disappeared in 1975, so the plans were realized nearly according to schedule. In addition, this committee compared the use of locomotive-hauled trains, EMU trains, and DMU trains. The report admitted, "Making precise, quantitative economic comparisons of all-EMU trains with electric locomotive-hauled trains is extremely difficult because there are many elements that tend to vary and are difficult to assess." Even so, the report continued, "If we take ease of operation and high speed into consideration, as a general rule, we have decided that all-EMU or DMU trains are the most profitable. Note, however, that if there are cases in which a locomotive can be used effectively for both freight and passengers, a train hauled by a locomotive may be more profitable."

Thus, the report concluded that not only the *Shinkansen* but also the conventional lines should make a big change to EMU and DMU trains. Once JNR set out in this direction, it would be difficult to turn back. The layout of stations and the railcar yards and inspection and repair facilities would be set up for EMU and DMU trains, and the technology and facilities of the companies that manufacture rolling stock would also center their efforts on EMU and DMU rolling stock.

European railways are overwhelmingly dependent on locomotives, and I asked the former General Manager of the Rolling Stock Department of SNCF why this was so.

"Both locomotive-hauled trains and all-EMU trains have their good points," was the answer that came back, "but I can't explain it in just a few words. I think it's probably a question of technological tradition, don't you think so?" Germany, too, formerly made a limited express EMU train called the ET403, but the Germans went back to using locomotives after making just a few of these EMU trains.

An executive of DB said, "Anyway, the prices are very high."

DB are going to try their hand at EMU trains once again. The ICE3, scheduled to go into service in 2000, will be their first all-EMU train in a long time. One of the high-level executive in DB engineers told me, "Actually, we've been plagued by a lot of problems with slipping when the locomotive-type ICEs reach high speeds." But I also got the impression that they wanted to have a varied fleet of trains because they were aiming to extend their lines into France and because they wanted to export rolling stock to China and other foreign countries.

3.6 DMU Limited Expresses Link All of Japan's Major Cities

With the success of the *Asakaze, Kodama,* and *Hatsukari,* the officials of JNR felt that the limited expresses were a hit.

Even if they were able to build the *Shinkansen* between Tokyo and Osaka, they wouldn't be able to meet the demands of the new era if they didn't create a network of high-speed trains on other lines. In October 1961, JNR instituted the most massive revision of its timetable since the end of World War II, increasing the number of limited express trains running to every part of the country. All at once, the number of round trips on the limited expresses rose from nine to twenty-six, as the limited expresses, which had formerly been special trains running in only a few lines, became a network that connected all the major cities. DMUs played the lead role in this new fleet of limited express trains.

At this point, not much progress had been made on the railway electrification. Of the major trunk lines, only the Tokaido Line was completely electrified. The Sanyo Line had been electrified only from Osaka to Okayama, and on the Tohoku Line, electrification work had just been completed as far as Sendai. If JNR wasn't going to use steam locomotives, it had no alternative but to use DMU trains. Furthermore, DMU trains had already proved their worth before the revision of the timetable.

DMU trains are noisier than EMU trains, and they also vibrate a lot. For that reason, the former common sence had been that they were suitable only for locals.

What undid the common sence was the *Nikko* DMU semi-express, which appeared on the scene in 1956.

JNR may have made it a semi-express because of a belief that it was impossible to adopt DMU trains for limited express or even ordinary express, but the *Nikko* showed his agility by running between Ueno and Nikko in just two hours.

One who sat up and took note of the *Nikko*'s success was Tadao Takano, Director of the Train Schedule Division, Train Operations Department of the Western Branch Office of JNR. Within two years, he had a DMU semi-express, the *Hikari,* running on a semicircular route around northern Kyushu from Hakata to Kokura, Beppu, and Kumamoto.

The Headquarter of JNR was vehemently opposed to this semi-express. The reason was that it ran faster than the limited express between Hakata and Kokura. However, Takano stepped up the operation of the *Hikari,* and from then on, one new DMU semi-express train after another went into service in Kyushu.

Takano was a confirmed speed fanatic, and when he became Director of the Train Operation Department at the Takasaki Railway Administration Bureau, he put into service a semi-express from Ueno to Maebashi. It was called the *Akagi,* and it was said to be an annoyance to its drivers. Next, he put another EMU semi-express called the *Haruna* into service between Ueno and Naganohara. Since the section from Shibukawa to Naganohara had not yet been electrified, the pantograph was lowered, and a steam locomotive hauled it. It could still run under those circumstances, but there was no power supply for the lights or opening and closing the doors. That's why a passenger car containing storage batteries was added to the train.

Fig. 3.5 *Hatsukari*, the first DMU limited express. *Photo* provided by the Railway Museum

The first DMU limited express to appear on the scene was the *Hatsukari* (Fig. 3.5) It had started as a passenger train hauled by a steam locomotive, but since it didn't use the special rolling stock designed for limited expresses, it was a slow limited express. It traveled via the Joban Line, which had more double-track sections than the Tohoku Line, but even so, north of Taira Station (present-day Iwaki), it was single-track all the way to Aomori.

The *Kodama* EMU limited express had already appeared on the Tokaido Line, and the *Asakaze* was formed of an entirely new type of passenger car, so in comparison, the *Hatsukari* didn't look as good as these two trains did, even though it was also a limited express. That's why it was decided to manufacture new DMUs for use with the *Hatsukari,* and they were completed in the fall of 1960. It was precisely in the fall of 1960 that an Asia-Africa Railway Conference was held in Tokyo under the leadership of President Sogo. He invited the participants and held a spectacular debut test running.

Things were fine up to this point. In the fall of 1960, the *Hatsukari* DMU limited express immediately began running on the same schedule as the locomotive-hauled passenger train of the same name, but the train experienced one accident after another. There were no catastrophic accidents, but since fires broke out during two of these accidents, there were huge repercussions. Newspapers carried large photographs of the burning *Hatsukari,* along with the punning headline, "This isn't the *Hatsukari* ('first wild goose'), but the *Gakkari* ('disappointment')."

At length, after the third fire in July 1961, President Sogo found himself forced to apologize. "We were in a bit too much of a hurry to produce the *Hatsukari,*" he said. "This string of accidents is absolutely inexcusable. However, the new type of limited express that will start running in October will be all right." JNR hurriedly set up an 'Ad Hoc *Hatsukari* Accident Fact-Finding Committee' and began looking into solutions to the problem, because the new DMU limited expresses that President Sogo had spoken of would soon be completed.

As I have mentioned before, the DMU limited expresses had been the flagship of the October 1961 schedule revision. For this reason, JNR had produced 127 of the new remodeled versions of the *Hatsukari* cars, and had set up new depots for DMU limited expresses at Hakodate in Hokkaido and Mukomachi in Kyoto, in addition to the depot at Oku in Tokyo.

Of these, the largest depot was the new one at Mukomachi in the suburbs of Kyoto. It housed 78 cars, mostly for trains that originated in Osaka, such as the *Hakucho* to Aomori, the *Kamome* to Nagasaki and Miyazaki, the *Midori* to Hakata, the *Heiwa* to Hiroshima, and the *Matsukaze* to Matsue. For this purpose, a new depot was being built in Mukomachi on the outskirt of Kyoto.

3.7   To the Mukomachi Train Operation Depot

When it came time to choose the person responsible for establishing the Mukomachi Operations Depot and keeping an eye on the trouble-prone DMU limited expresses, JNR somehow latched on to me. I had joined JNR just a little under five years before, and I was a 27-year-old novice.

At that time, these sorts of personnel practices were not at all unusual in JNR. Around our sixth year in the company, we were given experience as chiefs of depots, working right on the front lines, getting to know how the employees felt, and developing our talents as management executives. I think that these personnel practices may have arisen in the nineteenth century in imitation of the personnel practices for bureaucrats in the West.

It was a system that probably worked well enough in an era when there were no labor unions and the simple fact of being a university graduate gave a person significant authority, but things had turned a corner, and for that reason, I went through some rather painful situations, but it was also a valuable experience.

I was quite nervous as the time came to take up my post as chief of depot. Most of my seniors who had been through the experience said, "It's good to be a chief of depot. You're king of the hill." or "If a 'bachelor' goes to a difficult site, things are soon under control." These encouraging words were at odds with what I really felt, and some people even spoke to me in ways that seemed to be somewhere between encouragement and a threat. "The success of this fall's schedule change rests on your shoulders."

To be honest, I had no self-confidence at all. My head was full of such matters as what I would say to the employees when I called them all together soon after my arrival and made my opening remarks.

At the time I took up the job, the new Mukomachi Train Operations Depot was not yet in service. My official title was Chief of the Kyoto Passenger Car Depot.

The Mukomachi Train Operations Depot began operations in the fall of that year, with the Kyoto Passenger Car Depot as its parent organization and with the addition of the corps of locomotive technicians from the Umekoji Engine Depot and the engine specialists from the Kyoto Automobile Plant scheduled for abolition. This depot not only maintained the DMUs for the limited express; but it was also the foremost depot in Japan, the home base for drivers and more than 200 passenger cars.

In practice, this setup was a huge problem. First of all, the people from the passenger car depot didn't want to come to Mukomachi since it meant leaving a workplace where all the employees had gotten along like a family for many years and meant working together with unfamiliar people from the engine depot and the automobile plant. The traditional and beloved name of the Kyoto Passenger Car Depot will be disappeared. The employees had to move from their convenient location in front of Kyoto Station to Mukomachi, which was in the middle of nowhere. Under the surface of the seemingly splendid launch of the new limited express trains was a work site rife with unease and dissatisfaction.

There was a reason that I had been chosen for such a troublesome job. It was more than JNR's belief in a career path that was out of step with the times. There were also problems of disputes among factions.

The Technology area of JNR had several specialized departments, including Construction, Track Maintenance, Train Operations, Rolling Stock, Power, Signals & Communications, and Architecture, and each of them created its own faction. In particular, there were the endless complications of the power struggle between the Train Operations Department, which was in charge of maintaining the rolling stock, and the Rolling Stock and Mechanical Engineering Department. Within the Train Operations Department itself, the people in charge of locomotives, EMU trains, and passenger and freight cars each created their own factions.

It's only natural for such specialized groups to develop in human society, and it's not necessarily a bad thing. Factions can even be a source of energy as ventures get underway. Yet they often become a hindrance during withdrawal from an activity or major reforms of the organization. With employees drawn from three different groups, the Mukomachi Train Operations Depot would never become a cohesive unit, no matter who was in charge. That's why I, someone with no factional affiliation, had been picked out to supervise this site.

I received my baptism into this kind of factionalism during my first day on the job.

This was before the *Shinkansen* existed, and when I arrived on the *Tsubame* limited express to take up my position, ten or more employees in work uniforms came out to Kyoto Station to greet me. I nervously gave some sort of formal greeting. That evening, there was a welcome party for the new and old depot chiefs and the new and old deputy depot chiefs in a small restaurant near Kyoto Station.

Since the work site was undergoing the kind of reform seen only once in a hundred years, JNR had decided to put new people in all the positions of responsibility. It was a unique gathering. In addition to new and old depot chiefs and the new and old deputy depot chiefs, there were the officers of the local union and their parent organization, the union for passenger and freight car workers at the Osaka Railway Administration Bureau. Chairing the gathering were the assistant manager and the chief clerk of the Passenger and Freight Car Division of the Osaka Railway Administration Bureau.

An hour into the party, once the ice had been melted, the assistant manager began to speak slowly and deliberately in a thick Kyoto dialect. "Mr. Yamanouchi, you really have your work cut out for you. You have to break up the Kyoto Passenger Car Depot, and you have to build up Mukomachi, and make that accident-prone DMU limited express run properly, and arrange for consigning the railcar cleaning work to outside contractors, and there's still more. Because of the revision of the inspection regulations this fall, you have to restructure operations' work schemes. And you have to do all this in one year. We will back you up. We will, but from this day, you will be one of us Passenger and Freight Car people. If you go up against us, we'll get rid of you."

I had just arrived on the job, and here was a warning to become a member of the Passenger Car group and to work as an ally of the Passenger Car side on the work site.

The speaker, Minoru Matsuda, was a forceful man with a receding hairline, a bright, sharp glint in his eyes, and a threatening voice. With more real authority than the director of a section, he reigned over the world of passenger and freight cars in Osaka as the boss. People called him by nicknames, 'Baldy Matsu' or 'Sweety Matsu,' that were puns on his real name. These kinds of bosses still flourished in JNR at that time, and they could control both labor and management with a glance.

I received another baptism of fire on my third day. It was a kangaroo court.

Before I had become the depot chief, two depot employees had put up posters on the Blue Train. There had already been instances of people pasting up posters in JNR, but up to this point, pasting them up on the limited express had been taboo. An unusually strict punishment was handed down for having broken this unwritten law. Especially bad was the fact that one of the culprits was an important figure, the top union man in the Kyoto area.

I was summoned to the Railway Administaration Bureau and handed the notification of punishment, but I didn't know what to do after that. How was I supposed to get in touch with the person in question? Should I present the notification myself or have the assistant manager do it? Should I do it by phone or in person? How about my choice of words? After much anguished thought, I picked up the phone and said, "Come in at ten tomorrow morning, because I'm going to notify you of your punishment."

Well, they came all right. The next morning at ten, about twenty people, almost too many to fit into the depot chief's office, showed up wearing red armbands and red headbands. All of them looked angry, almost demonic.

I would like to inform you of your punishment.

Just a minute. There's no reason for me to be punished.

Well, now didn't you paste up posters on a limited express?

What proof do you have that I pasted up posters? Were you watching when it happened?

No, I wasn't watching, because it happened before I came on the job.

What are you talking about? What's the idea of punishing me for something that you didn't see?

Well, there are people who were able to establish it as a fact. The organization has a grasp of the facts.

You're young, but you're also the depot chief. If we were talking about a family, you'd be the father. A father should protect his children, right? Did you go to the Railway Administration Bureau and protect me? Go tell them, 'I didn't see it happen, so I won't accept this punishment.

Well, the facts have been properly eye-witnessed.

If the facts have been eye-witnessed, what's the name of the guy. . .?

No, that...

If you talk like that, you're not fit for the job of depot chief. No one will follow you. Get yourself back to Tokyo right now.

We continued arguing along these lines for more than an hour. At one point, the most hefty-looking man stood by the side of my desk, pounding on it each time I spoke and kicking my chair as he shouted into my ear, "What kind of attitude is this?".

That's when the chorus of complaints changed its tune from opposition to punishment to criticism of my attitude.

What do you mean, talking like that? What's the idea of sitting with your legs crossed when one of your beloved subordinates is being punished?

I did have my legs crossed, as was my habit. But I thought that it would be unmanly to uncross my legs while this mob was ganging up on me, so I just kept them crossed. After a while, my legs became numb.

"All right, here's what I'll do," I said.

All at once, everyone fell silent.

I'll apologize for my attitude. But I can't rescind the punishment.

The other side must have thought that this was their chance.

What an idiot. We'll be back tomorrow.

They left, one after the other.

Frankly, it was a terrible shock. I had never thought anything like this could happen to me without warning.

At that time, most work sites were still peaceful, which is why so many of my senior colleagues told me "It's good to be a chief of depot. You're king of the hill." However, Kyoto had already begun to enter the era of change. Maybe my attitude was inappropriate. Twenty years after the fact, I ran into the man who had pounded and kicked over my desk at JNR Headquarter in Tokyo.

Hey, remember me?

I sure do. I never saw such an uppity guy as you in my whole life.

Perhaps I was just insisting on my own opinion so as not to be made a fool of.

That night, I hardly slept a wink. How would I answer them if they came after me again the next day? Even thinking about it threw me into confusion. I was in my room alone, and even the white walls seemed threatening. I ran out into the streets of Kyoto and wandered aimlessly through the night until dawn.

The next morning as I waited, resigned to whatever might happen, a union officer came alone.

With that, he took the notification of punishment from me and left.

Rather than feeling relieved, I felt that something was off kilter. When I thought about it later, I wondered if *Baldy Matsu* had talked the union into accepting the punishment. He had the kind of power that allowed him to tell the union not to destroy the young depot chief.

3.8   The Lonely Chief of Depot

Compared to some of the lynch mob tactics that were used at JNR work sites later, this kangaroo court may hardly be worth mentioning. Beginning in the 1970s, terrible mob-like tactics prevailed at most JNR work sites under the name of 'the work site negotiation system.' It was really group sadism passing itself off as a labor movement. I had workers gang up on me any number of times under the guise of collective bargaining, but this first experience was the most intense. It was, of course, my first such experience, and the position of the depot chief is a lonely one.

Since the site where I was working was the Passenger Car Depot, all the workers belonged to the JNR Workers Union, but only a few of them were really zealous about the labor movement, and even they showed no signs of resistance during their everyday work. The majority of the people were extremely serious and dedicated, and the workplace was like a big family.

The first problem I ran into here was the problem of education. All the people had worked on passenger coaches throughout their working lives, and no one knew anything about DMUs. During half a year, I had to educate these people to be the main group responsible for taking care of the accident-prone, complexly structured DMUs. First, I would have to send them to a place called the Education and Training Center for some basic education.

I chose the most clearly outstanding workers and tried persuading them to take the training.

But all of them refused.

> I've never worked on anything but passenger coaches, so that's enough for me. I don't even want to go to Mukomachi.
>
>> Then we won't be able to go through with this schedule revision. I'm depending on you, please. If you don't go, the new Mukomachi Depot will be ruled by the people from the Engine Depot and the workshop. In the meantime, the Passenger Car Depot is going to shut down.

I tried threats and persuasion, but nothing worked, and the time limit was fast approaching. Feeling that I had to do something, I decided to teach two classes per week on diesel engines myself, and I put a notice on the bulletin board informing the workers of these classes. I made all the teaching materials myself. And to my surprise, several dozen people showed up.

It may be that the workers went for training because they had gradually realized that the changeover was inevitable. And yet, perhaps, I regret to say, it wasn't due to my efforts but to *Baldy Matsu*'s power.

One of the workers asked, "I'll go along with what you want, sir, but aren't DMU trains going to be replaced by EMU trains?".

> It'll be fine, I replied. DMU will be all right, at least during your working years.

About twenty years after that, DMU limited expresses disappeared from the Mukomachi Depot. When the Aomori-bound *Hakucho* limited express was changed over to an EMU train, there was a ceremonial last run for the DMU Type. I was invited to ride along, and I was also delighted to meet up with that man, who was now retired.

At the beginning of August, shiny new DMUs for the limited express were rolled into the Mukomachi Depot one after another. But the depot itself was still under construction, and just a few brand new tracks laid down on an expanse of earth that has been reclaimed from the rice paddies. As we worked under these conditions, strict orders were issued from the Headquarter, telling us to conduct test runs from Osaka to Aomori or to Nagasaki and Miyazaki every day without fail.

In any case, since the *Hatsukari* continued to break down, we tried to run it as much as possible before the schedule revision to get a grasp of what kinds of breakdowns were occurring and how we should fix them.

As might be expected, we were extremely unprepared in many respects. We had blundered in too many ways, and while the tracks, at least, were partly completed, the buildings were still under construction, some still just wooden scaffolding. The one room that we could more or less use served as my office and as the on-duty managers' room. There were no rooms for the people who inspected and repaired the rolling stock, and of course, there were no lockers. Instead of having a crew member's room, we brought in two sleeping cars to use as temporary changing rooms.

Since the workers would end up covered with oil after repairing and maintaining the DMUs, bathing facilities would be essential. We somehow managed to get a bathing room built ahead of schedule, but the boiler didn't arrive on time.

At that point, we sent for an old-style steam locomotive and used it as a temporary boiler.

When I ordered the first contingent out to the new depot, the head of the local union came to my office alone.

'Chief,' he said in his Kyoto dialect, "It's no good. We aren't going. If you'd go and see Mukomachi for yourself, you'd see that there's nothing there. No rooms. No baths. No toilets. No food. What are we supposed to do? We aren't dogs, sir. If we were dogs, you could grab us by the scruff of the neck and take us there, but we just won't tell everyone that they have to go."

> Look, I'm sorry. I need you to do this. Just put up with it for a little while. It's for the good of JNR.

The head of the local union grimaced briefly and left. The next day, the first contingent left for Mukomachi.

The test run was exhausting. Because of breakdowns on successive dates, the train returned late. Sometimes the engines on half the thirteen cars stopped working. Since the Headquarter had ordered us to do test runs every day, no matter what, the inspection gangs spent sleepless nights fixing one thing or another. The pattern of breakdowns and late returns was repeated over and over.

Kyozo Kondo, who was principal engineer of JNR Rolling Stock Design Office, the office mostly responsible for the design of DMUs, recalls those days in this way:

> Until then, DMUs had run mostly on lines with few passengers, and the trainsets tended to be short. Since there wasn't enough horsepower from one engine, two engines were installed in each car. For example, each *Hatsukari* run had nine cars, so each train had 17 engines.

Only the dining car had a single engine. With 17 engines, the accident rate—well, if you have one engine in a one-car train, the probability of an accident is one in 10,000. If you have 17 engines, that probability is multiplied 17 times. I could just feel it in my bones. If you think in terms of a family, and it's certain that if you have 17 children and one of them will have a cold, it's a lot of trouble for the father and mother. We'd hear someone say, "Uh-oh, another engine stopped," and then we'd hurry to Oku depot. We went to a lot of trouble to find out the cause.

At the beginning of September, we put up a signboard reading 'Mukomachi Train Operations Depot' at the new depot, which still looked like a construction site, and I also moved my office from Kyoto to Mukomachi. That being said, we still had only one usable room (Fig. 3.6).

A few days later, the most doctrinaire labor union member at my work site came in with a complaint.

Sir, what's going on here? It's a violation of the Labor Standards Law. Even though we don't have concluded Article 36, you're still overworking us.

By 'Article 36,' he meant Labor Standards Law Article 36, which stated that an agreement with the labor union was needed before workers could be made to work overtime. But at that time, JNR unions refused to conclude such agreements. They had no right to strike, and for this reason, they often used refusal to conclude an Article 36 agreement as a legal resistance measure. At this time, they were also using this means to express their opposition to the schedule change and to rationalization.

No, I said, I'm not ordering anyone to work overtime. Everyone's working voluntarily.

Fig. 3.6 A group photo at Mukomachi Depot in front of the trouble-prone DMU limited express. *Photo* provided by the Railway Museum

I don't care what you say. It's no good.

Even with that, there was no trouble. The sector workers pitched in and worked at night without sleeping, and this person didn't come after me anymore.

As the revision of the timetable approached, one final crisis awaited us.

Because JNR Workers Union was opposed to rationalization, it began holding work site assemblies during working hours, and it was decided that the Mukomachi Train Operations Depot would be the base for this movement. This was also a battle tactic that JNR labor unions, denied the right to strike, had thought up. At the beginning of their shift, the workers showed up, but they held work site assemblies instead of working. Since this kind of action was a sort of strike, it was considered punishable. Yet if the assembly ended within 29 minutes, the workers were not subject to wage cuts.

The day before the work site assembly, the head of the local union once again showed up unannounced.

Sir, what are you going to do, huh?

Is there anything I *can* do? If you're going to do it, just do it.

No good. That's no good.

The head of the local union was named Shinnosuke Shimizu, and he was about fifty years old. He was a stout and dignified yet sedate person, and he seemed distant from the real fighters in the labor movement. He was popular with the other employees, and he had the disposition of a boss. He smiled in a way that gave the impression that he could tell what other people were thinking. Like a typical native of Kyoto, he never stated his feelings clearly. He often looked at me, an outspoken type from the Tokyo area, and said, "Sir, because you talk like that, the things that need to be settled don't get settled" or "Don't you understand what I'm saying? I'm leaving."

However, this time before the meeting, he had a serious expression on his face. He said that there was something he wanted to discuss with me alone, but we were being observed by left-wing union members. He returned to his home in Sakamoto, Shiga prefecture, and I returned to my apartment in the company housing block. After taking a bath, I went by taxi to Okoto Hot Spring Resort where I spoke with Shimizu.

Executives from the Railway Administration Bureau also showed up at the work site assembly, which began in a ceremonious manner at one o'clock on the afternoon of the next day. It was during working hours, but the assembly continued, and after about ten minutes had passed, I went by myself to the place where it was being held and ordered them to stop this assembly immediately and go to work.

The assembly soon ended, and I went over the facts with the union members in my office.

One of the officials from the Labor Division stood up and whispered in my ear, "Let's just pretend this never happened."

Shimizu and the vice-chairman of the upper-level union organization came into my office, looking tense. The vice-chairman quietly began to speak.

Sir, do you have any idea how long the work site assembly was?

I hear that you all went back to work at the appointed time.

I'm told that afterward, Shimizu was seen on the work site looking very happy.

Two days before the schedule revision, the train drivers newly assigned to the Mukomachi Train Operations Depot invited me to have dinner with them. Since no drivers had been assigned to the Passenger Car Depot, these were the first ones I was supposed to supervise, and they sent a representative to issue the invitation.

Sir, once the schedule is revised, all of us drivers won't be able to get together. Please join us.

That evening, even I, someone who doesn't handle liquor well, drank so much that I got sick.

Finally, the day of the schedule revision arrived. The night before, the main people and I, of course, all stayed overnight at the work site, taking scrupulous care to ensure that the DMUs were ready. All the trains that would depart from Mukomachi on that day— the *Hakucho* to Ueno and Aomori, the *Kamome* to Nagasaki and Miyazaki, the *Heiwa* to Hiroshima, and the *Matsukaze* to Matsue—were lined up. Operation of the *Midori* to Hakata had been suspended for the time being. Since the new type of limited express broke down too often, there were never enough parts, and since we took spare parts from spare rolling stock, we ran out of spare cars. In effect, the spare cars had served as parts warehouses.

3.9   The Vaunted *Matsukaze* and *Hakucho* Break Down

I rode along in the driver's cabin of the *Hakucho,* the train with the longest run. The inspection gang, who had not slept at all the previous night, saw me off by raising their hammers in a salute, and the train began deadheading toward Osaka.

After we passed through the Miyahara Marshalling Yard, had arrived at Osaka, and were waiting to start up again, the assistant station master came running.

The orientation of the train is reversed! You have the dining car where the first-class car should be.

Oh, no! They were back to front. When the *Hakucho* arrived at Osaka from Aomori after the beginning of passenger service, it would first be heading west, but at the Miyahara Marshalling Yard, it would turn east to return to the Mukomachi Depot. The arrangement of the cars when they arrived at Mukomachi was the opposite of the order they were supposed to be in during the run from Aomori to Osaka. However, during the test runs, there had been no need to go to Osaka, so we had returned directly to Kyoto. That's why the cars were in the wrong order. We had been so busy that no one had noticed (Fig. 3.7).

The car number on the first-class passengers' tickets were now wrong, but we somehow got them into the right seats and started off. The painstakingly prepared

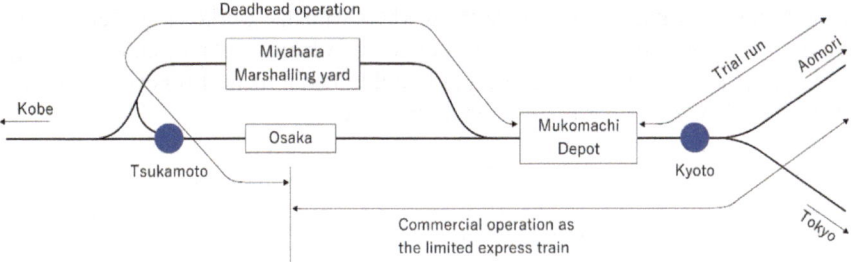

Fig. 3.7 Schematic layout of the Osaka area with turnaround points of the DMU limited express *Hakucho*

engines ran smoothly and without any trouble as far as Tsuruga. This is where the Mukomachi-based driver was replaced. I watched the limited express take off for the Hokuriku Tunnel, said hello to my colleagues in the Tsuruga Engine Depot, and returned to Mukomachi.

When I returned, a manager came running, "Sir, it's bad news. The *Matsukaze* broke down."

The bearings in the last car of the *Matsukaze* had overheated, so that car had become detached at Fukuchiyama, and the train had left for Matsue an hour late. I was told that the mayor of Fukuchiyama and a brass band, who had come to the station to celebrate the arrival of the first limited express, were left waiting on the platform for an hour. Since there was no driver's cab for the return trip to Osaka, one car with a driver's cab was hurriedly sent out as a deadheaded trip.

The DMU limited express, which seemed to be running properly in spite of everything, finally caused great confusion on October 21.

The Osaka-bound *Hakucho* broke down completely at Tsuruga and had to limp into Osaka hauled by a C57 steam locomotive.

The cause was a breakdown of its electrical circuits. This was the kind of problem that had plagued us most during the test runs, and once the breakdown occurred, not only did all the electrical circuits and the engine shut down, but it was difficult for us to figure out what was wrong and pinpoint the cause. The newspapers carried large photos of the *Hakucho*[5] being hauled by the locomotive, and they made jokes about 'The Dying Swan.'

Soon an aide from JNR Headquarters, someone a few years older than I, came to Mukomachi.

Everyone from the president on down told you absolutely not to let any accidents happen, he scolded, and look what happened. You're worthless.

I lost my temper. It wasn't as if I didn't have any second thoughts, but JNR had built some accident-prone rolling stock and forced it upon this work site and brought together a group of novices and trained them. On top of that, the construction of the depot hadn't been completed on time, and there hadn't been enough parts. Foremost

---

[5] Swan in Japanese.

in my mind was what I would say to the workers who had worked together with me, staying up nights, despite pressure and opposition from the labor union. I reacted fiercely, but tears came to my eyes, and I couldn't stop them. I'll never forget how angry, humiliated, and frustrated I felt at that time.

If you count my time at the Kyoto Passenger Car Depot, I was at the Mukomachi Depot for a mere 15 months before I was relieved of my duties and transferred to the Headquarter.

The day I left, almost all the employees came to see me off.

As I left, several people ran after the automobile I was riding in, pounding on it.

Chapter 4
The Struggle of the Technicians

The author was in charge of the tests for switching to adhesion operation on the 66.7‰ apt section at the JNR Headquarters. It was a very challenging project where a coupler broke during the course of the test.

At the time, JNR was aiming to increase the speed of express trains, but there was strong opposition from the infrastructure side, which argued that this would cause significant track damage. The author sought a compromise by talking with members of the infrastructure side to learn of their concerns.

When trains run at high speeds, the phenomenon of hunting, in which the railcars sway laterally, occurs. After the defeat in the war, engineers who relocated from the military brought a new wave of technology to this problem.

In this chapter, the efforts of engineers in various fields of railways on the eve of the *Shinkansen*'s debut are introduced along with episodes from that time.

4.1 The Performance Test between Yokokawa and Karuizawa

My new job at the Headquarters was planning for speed, but my formal title was 'Assistant Chief, Train Schedule Division, Train Operations Bureau.' Each time new locomotives or EMUs appeared one after another, my job was to assess its performance in terms of how many minutes it took for it to run from one station to another. This was basic data for creating the train schedules. In one sense, it was a thankless task, but since I was in charge of test runs of new rolling stock, it was quite interesting in a technological sense, and I learned during my time in that position.

Yet I encountered two thorny problems.

The first was confrontations with my subordinates. The entire staff of the department I managed was a mere five people. Of the over 400,000 employees JNR had at the time, the ones who came to the Headquarters were the cream of the crop,

© Japan Railway Technical Service 2024
S. Yamanouchi, *If there were no Shinkansen*,
https://doi.org/10.1007/978-981-99-8890-7_4

strong personalities, and as a result, the Headquarters was quite a collection of self-confident types. When I arrived at my new position and gave my self-introduction and remarks, everyone greeted me in response, but soon after that, my top-ranking subordinate blithely turned his back to me and began speaking loudly about his work.

The Vernier control of the next new model of the locomotive using a magnetic amplifier...

Having been preoccupied with breakdowns in DMUs and labor problems until the previous day, I didn't have the faintest idea what he was talking about.

His turned back seemed to be a silent message: "You may be my superior starting today, but I bet you don't understand anything." It was mortifying, but since I really didn't understand anything, there was nothing I could do but put up with it.

About a month later, there was a meeting of everyone in charge of speed planning nationwide. First, I gave the opening remarks in my capacity as an Assistant Chief from the Headquarters. When I was finished, that same employee stood up suddenly and blurted out, "What the Assistant Chief just said is wrong. Here's the real story."

This is when I lost my temper. It was true that I didn't know as much as I should, but I thought, "What a bastard."

After that, I went to the office library, borrowed explanatory material about the new models of rolling stock, and studied them every evening at home. I didn't have time to do this during the day time, and besides, I didn't want to be seen boning up on job-related material.

During that time, I had many strong disagreements with that particular subordinate concerning technical matters. If I lost the argument, I also lost my temper. If I won, he turned his back on me. He was older than I, so there was nothing I could do about it.

That was about the time that large-scale test runs between Yokokawa and Karuizawa began.

At the same time that we were changing that section from single track to double track, we were also switching from an Abt-type system with gears to a new type of locomotive. In addition, since the section had a 67‰ gradient, it would be terrible if anything went wrong, and we would have to take extra care in running the tests. Could the new locomotive run with the planned number of freight cars attached? Would it deliver the planned performance? Would it slip when it rained? Were the brakes all right? What would happen if there were a power outage during the run? In the unlikely event that the locomotive broke down, what then?

The second problem for me was the question of which department would be in charge of the test runs for the new locomotive.

One of the inherent weaknesses in JNR was sectionalism. Even if you were aware of this, it was hard to escape the tendency once you were part of the organization. The reason was that both the evaluation of your work and the course of your life would be determined within this little world. I believe that the major surgery that JNR underwent when it was reformed was absolutely essential.

Two bureaus of JNR were in charge of operating and maintaining the rolling stock, the Rolling Stock and Mechanical Engineering Bureau and the Train Operations Bureau. The design and planning of rolling stock and major repairs at the workshop were under the supervision of the Rolling Stock and Mechanical Engineering Bureau, while the planning of daily operations and inspection at the EMU depots was under the supervision of the Train Operations Bureau. These two bureaus were notorious within JNR for being on bad terms. In a word, they were involved in a power struggle. For the most parts, their roles were defined, but disputes sometimes arose in areas that were on the borderline of their respective turfs. One of these areas was the performance testing of new rolling stock.

The Rolling Stock and Mechanical Engineering Bureau, naturally enough, insisted that the performance testing of newly produced rolling stock was its responsibility. One couldn't argue with that, but the Train Operations Bureau insisted that it was the bureau that used the rolling stock, and therefore, its staff needed to carry out performance tests of the items that they themselves needed to know. That's what I insisted. Both sides argued, and in the end, both sides conducted separate performance tests. I thought that the Rolling Stock and Mechanical Engineering Bureau should take the main role, and that both sides should work together, in fact, but if I had said that, my reputation would have suffered, which would have delighted my top-ranking subordinate. He kept quiet, but the expression on his face was a silent challenge to my authority.

The Rolling Stock and Mechanical Engineering Bureau's tests stretched out over two months, and when they had ended, the Train Operations Bureau's tests began. The Rolling Stock and Mechanical Engineering Bureau's tests were centered on the question of whether the new locomotive could deliver the planned performance, but the Train Operations Bureau's centered on a different set of questions. Could the locomotive run as fast as planned? Could it run according to schedule? Could it start up again if it stalled on a gradient? Would the emergency and safety equipment work well in case of a power outage or breakdown?

During these tests, another dispute broke out between me and my top-ranking subordinate.

It arose over the problem of the locomotive slipping when it started up after stopping halfway up a gradient. Everything was fine when the locomotive was running, but once it stopped, it met with tremendous resistance when it started up again. If the power on the locomotive was increased, it would sometimes slip and be unable to move. The most difficult times were when a light rain was beginning to fall. We weren't favored with rain during the tests, so we sprinkled water on the tracks on the gradient. This was one of the biggest problems in determining whether use of the new type of trains operating between Yokokawa and Karuizawa would work out or not.

The EF63 locomotive had been devised so that if one of its driving wheels began to slip, the electric power to that driving wheel was quickly weakened. I think my dispute with my subordinate was about the technology of the process by which this slipping was halted. In the course of the argument, this subordinate flew into a rage, ran out of the test car, and went off somewhere.

Startled by his outburst, I was waiting for him when he returned, dragging a researcher from the Railway Technical Research Institute by the arm. Our argument started up again. The researcher listened silently to both sides and finally said, "The Assistant Chief is right."

After that incident, my top-ranking subordinate listened to what I said. He was an enthusiastic and knowledgeable person, and I could even say that I was able to get my job done because he was around.

In any case, this Yokokawa–Karuizawa test ran up against a huge, unforeseen problem. But before I tell you about it, I would like to provide a simple explanation of the history of this section.

4.2 The Steep Gradient at Usui Pass

This section was first opened in 1893, just four years after the opening of the entire Tokaido Line. Since there was a possibility that the Tokaido Line, which ran close to the coast, would be destroyed in the event of a war with a foreign country, construction moved forward on the Nakasendo Line, which ran through the mountains, as a second line connecting Tokyo and Osaka. However, a difficult barrier known as Usui Pass laid on the way. In the mere 11 km between Yokokawa and Karuizawa, the train had to climb 500 m. That was how the steep 67‰ gradient railway came to be.

A gradient of this degree is by no means unusual for a road, but for a railway, it is extremely exceptional. The Hakone Tozan Railway runs on a steep gradient with an 80‰, but it is literally a 'mountain-climbing' (*tozan*) railway exclusively for EMU trains. For the trunk lines that carry long-distance trains and freight trains, a 33‰ gradient is the limit. The Yamagata *Shinkansen* between Fukushima and Yonezawa is one example of a line with a 33‰ gradient.

Since the gradient in this case was twice that steep, four rack-and-pinion-type locomotives were imported from Germany and began operating on that section. This was the birth of the Abt system railway. At the time it opened, there were only five daylight roundtrips between Yokokawa and Karuizawa, each of which took an hour and eighteen minutes.

On this section, locomotives were usually linked to the rear end of the train. This was to prevent the passenger cars from careening wildly down the gradient if the couplings happened to come loose.

In addition, since there were 26 tunnels on this section, the smoke produced by the locomotive became a major problem and measures were taken in response. Not only did attaching the locomotive back to front while climbing the gradient prevent the smoke from entering the driver's cab; there was also an exhaust smoke curtain installed at the entrance to the tunnel.

When a train approached a tunnel, the signalman opened the curtain and closed it once the train was inside. It was to prevent the smoke from the locomotive from getting into the driver's cab. Accidents occurred when the signalman forgot to open the curtain and the train tore through it.

Passenger coaches and freight cars did not have the brake equipment at that time. Only the caboose, equipped with brakes, was coupled to the train and was responsible for all braking, so the situation was rather frightening.

In 1911, at the end of the Meiji Period, electrification of this section was completed, and electric locomotives, the first ones put into service in Japan, replaced steam locomotives in playing the main role at the pass. EMU trains were already running on what is now the Chuo Line, but there still were no other electric locomotives running, even on the Tokaido Line. At first, Japan used locomotives imported from Germany and Switzerland, but in 1919, the ED40 appeared on the scene. It was the first electric locomotive produced in Japan, and the Yokokawa-Karuizawa section was the birthplace of Japanese electric railway technology (Fig. 4.1).

Under the Abt system, one locomotive was attached at the front and three at the rear when climbing the gradient in the direction of Karuizawa. In cases when two or more locomotives were attached to one train, one of them was designated as the 'main engine locomotive.' This was the one that controlled the operation of the train, and the assisting locomotives had to follow its directions.

Usually, the locomotive in front was designated the main engine locomotive. However, when climbing Usui Pass, the locomotive at the very back end was the main one. Only from the rear locomotive was it possible to observe whether all the gears had meshed when the train entered the rack-equipped section. Since the most

Fig. 4.1 Abt-type electric locomotive ED42 (succeeding model of ED40), which used the rack-and-pinion system for steep gradients. At Yokokawa-Karuizawa on the Shinetsu main line. *Photo* Provided by the Railway Museum

dreaded possibility was that the train might roll down the gradient out of control, the rear locomotive was chosen to bear the responsibility of holding the train in place.

In those days, there were no wireless or remote controls, and the four engineers, each in their respective locomotives, used their whistles as signals. The signal indicating that the current was being turned on and that the electric motor was starting up was one short blast. In the opposite situation, when the current was being turned off and the train was going to start coasting, the signal was one long and two short blasts. In the old days, when the steam locomotives on the Shinetsu Line were replaced and passengers and crew were liberated from smoke, climbing among the mountains at a relaxed pace as the four whistles signaled one another was a most atmospheric experience.

At present, it is possible to go from Tokyo to Karuizawa in a little more than an hour on the *Shinkansen*, but in those days, there was only an ordinary express, and it took three hours.

The era of the Abt system lasted for 70 years, and there were two major accidents. The first occurred on July 13, 1901, when a combined freight and passenger train that had climbed the pass was just about to enter the last tunnel before Karuizawa. The boiler of the locomotive exploded, and steam poured out into the driver's cab. The brakeman jumped into action and held tightly to the brake in the caboose, so the train was able to stop after sliding backward for about two km, but Moori, Chief Engineer of the Nippon Railway (this line was still part of a private railway at this time), and his son jumped out of the runaway train and were killed.

The other major accident occurred on March 7, 1918. A locomotive on a freight train climbing from Yokokawa to Karuizawa broke down inside the 21st tunnel on the route. The engineer of the main locomotive consulted with the engineers of the assisting locomotives, and they decided to try starting up again. But the train suddenly began running backward, and although they frantically applied the handbrake, the handbrake that shut down the electric motor, the generator brake, and every other brake they had, the train just kept accelerating. The train flew into a siding at Kumanotaira Station at high speed and crashed into a stone wall. All four railway employees aboard were killed.

On this kind of a steep gradient, a runaway train can accelerate in no time at all and become unstoppable. Even so, it's amazing to think that the line ended up having only two accidents of this type in 70 years. Furthermore, in the early days, the passenger or freight cars, hauled by steam locomotives, had no brakes at all.

The Abt system was a safe and sure system for climbing steep gradients, and an atmospheric one as well. *Kamameshi*, a new specialty, the risotto-like dish, was also created by taking advantage of the locomotive changeover time.

However, any way you look at it, the Abt system was slow. A mere twelve kilometers took 46 minutes, and the average speed was 16 km/h. That wasn't all. Only ten passenger coaches could be attached, and since there was only one track, only 30 roundtrips could be run on the line each day.

In the latter half of the 1950s, a plan was proposed to both electrify the Shinetsu Line and make that section double track. By electrifying the entire line from Ueno to Nagano, we wanted to make it possible to run the whole trip with the same locomotive and put EMU trains into service on the line. We also wanted to increase the speed. The Abt system was unsuited for these purposes. For many years, it had served to maintain safety through the pass, but now, ironically, it had become a hindrance.

4.3 The Replacement for the Abt System

The Abt system facilities were old and worn out, and the time was coming when they would need to be replaced. Furthermore, the tunnels between Yokokawa and Karuizawa were too small to let either an EMU train or an electric locomotive with pantographs pass through.

The Abt system locomotives used specifically for this section did not use a pantograph but took their electric power from a third rail that ran alongside the track. This is the so-called third-rail collector system used on some subways, and for reasons of safety, the voltage was only 600 V, lower than in other sections. At the British end of the route, the *Eurostar,* the high-speed limited express linking London and Paris, still runs on the third-rail collector system.[1]

But would it be possible to climb this steep gradient without rack-and-pinion assistance? As I mentioned before, the Hakone Tozan Railway is able to climb an even steeper gradient of 80‰ without a rack-and-pinion system. In France, there is even a railway that climbs a 90‰ gradient without such assistance. And yet, since both of these lines are used only for short trains, conditions are different from those at Usui Pass, where both express trains and freight trains ran. JNR compared and studied six plans for making this section double track (Fig. 4.2 and Table 4.1).

The first plan proposed a relatively gentle 25‰ gradient, which is nothing unusual for mountainous sections. This was the easiest option from the viewpoint of the person driving the train, but it would be necessary to take quite a roundabout route to achieve it. Since this would increase the distance, it would not only lead to higher construction costs but also lengthen travel time.

The 33‰ plan involved creating a somewhat steeper gradient, but it did not present any technological problems either. The 50‰ plan called for building a gradient that JNR had no experience with, but it would probably be possible to run trains without rack-and-pinion assistance. The final plan was to add tracks parallel to an existing line with a 67‰ gradient. A new line would be built with large tunnels parallel to the existing Abt system line. When the new line opened, the Abt line would go out of service for a year, during which the Abt system facilities would be removed and the tunnels enlarged. After various deliberations, JNR decided on the double-track plan to add another track to the existing track because construction costs would be low and travel time on the new line would be short. The question, when it came right

---

[1] In November 2007, with the opening of the new HS1 high-speed line with an overhead contact wire system in the UK, *Eurostar* began operating via this line, thus reducing the travel time between London and the Continent by 20 minutes.

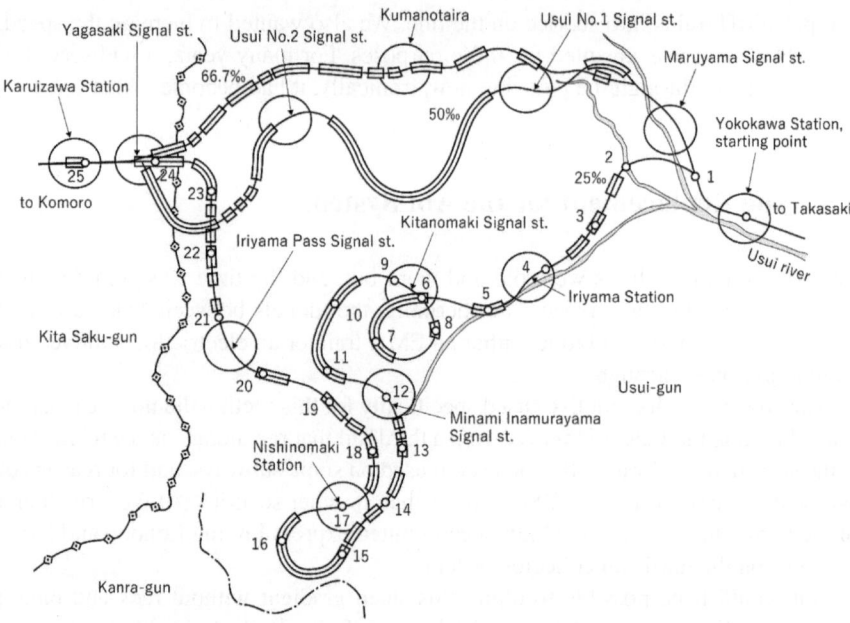

Fig. 4.2 Comparison of the construction routes for the 25‰ plan, the 50‰ plan, and the 66.7‰ plan. *Source* 'Transport Engineering' Kotsu Kyoryokukai Nov 1959

Table 4.1 Plans for improving Usui Pass

	Track length (km)	Construction expense (Billion yen)	Construction period	Maximum number of trips	Notes
25‰ detour plan	25.2	7.1	2 years and 9 months	240	
25‰ plan with 15 km of tunnels (single track)	23.0	5.0	5 years	75	Karuizawa station would be underground
25‰ plan via Naganohara line (single track)	–	6.5	5 years	60	60 km longer than 66.7‰ plan
33‰ plan	19.5	5.9	3 years	240	Could operate EMU trains
50‰ plan	13.9	5.6	4 years	240	
66.7‰ plan	11.2	3.6	2 years and 6 months	200	

Source 'Transport Engineering' Kotsu Kyoryokukai Nov 1959. Simplified by the Author

down to it, was whether a train could run safely without rack-and-pinion assistance on this steep section.

At just about that time, practical applications of alternating current(AC) led to theoretical analyses of slipping in electric locomotives, and there were new advances in control technology for them. The effect was that it seemed possible for trains to run on a 67‰ gradient without a rack-and-pinion system. This is precisely the question that my top-ranking subordinate and I argued about in the test car.

The next problem was the brakes.

As past accidents reveal, brakes for going downhill are a more important problem than power for going uphill on such a steep gradient. Normal friction brakes, with brake shoes attached, overheat and burn out if they are applied continuously. It was therefore decided to attach four kinds of brakes: electric brakes, electromagnetic rail brakes, and hand brakes, in addition to air brakes.

Electromagnetic rail brakes require installing a large electromagnet in the locomotive, and they are set up so that the electromagnet is attracted to the rail when the brake is applied.

The big argument we had about brakes was whether to adopt regenerative brakes. When electric brakes are applied, they use the electric motor of the locomotive or EMUs as a generator. When electricity is generated, a strong force works in the direction to stop the train, so it is used as a brake. However, when the train slows down, this force becomes extremely weak, so in the end, air brakes are essential for stopping the train.

In most electric brakes, the electricity generated is sent to a resistor and given off as heat. This is the reason that you will sometimes feel a warm breeze when an EMU train comes into a station.

But this is quite a waste of energy. If you take the electricity generated and instead of giving it off as heat, send it through the pantograph back to the overhead contact wires, it can be used to power other trains. This is the principle behind regenerative brakes, which are currently being used on the intra-city trains of the Yamanote and Keihin Tohoku Lines in Tokyo and on new *Shinkansen* EMU trains.

This technology is attractive for its rationality. It would be particularly effective on a section like Yokokawa–Karuizawa, where electricity could be continuously generated from the steep gradient. Chief Engineer Hideo Shima strongly urged the adoption of regenerative brakes for the new electric locomotives that were being built especially for this section.

However, that particular technology was not sufficiently developed at that time, and it would have been terrible if, by chance, the brakes had failed on this section. With just the air brakes, it would be very dangerous unless the speed had already been reduced.

That's why the executive engineers visited Chief Engineer Shima to explain that they were worried about using regenerative brakes. Shima was often described as being like a typical British gentleman, soft-spoken and dignified. As the executives talked about the various problems associated with regenerative brakes, he just made a series of non-committal comments:

That's right.

It's just as you say.

That may be so.

Finally he said,

I can really see what you mean.

By the time the explanation was over, the executives thought that Shima had accepted their point of view, but several days later, he issued directions to use regenerative brakes definitely.

What was going on here? In fact, Shima's "I see what you mean" meant nothing more than "I know what you're trying to say" and did not mean that he himself accepted the other person's point of view. It became widely known within JNR that one should be careful about interpreting Shima's "I see what you mean."

Work on doubling the track began in April 1961, and two of the new-style loco-motives were completed in June of the following year. By that time a mere two kilometers of track were ready around the Maruyama Signal Station near Yokokawa Station, and there the first series of performance tests began.

4.4   Conversion to Double Tracks and the New-Type Locomotives

The locomotives were two: the EF62, for running between Ueno and Nagano, and the EF63, assisting engine locomotive especially for this section. There was nothing particularly unusual about the EF62, but the EF63 was a special locomotive with all sorts of equipment attached.

First of all, it was heavy. The weight on each axle in a typical locomotive was at most 16 tons, but in the EF63, it was 18 tons. Metal weights were fitted to it to make it heavy, because the heavier the axle weight, the less likely the locomotive was to slip. It had a piece of equipment known as a slipping detector, set up so that if one driven axle began slipping, the current sent to the electric motor would be weakened, thus automatically stopping the slipping.

It also had a special small axle for measuring speed. In typical electric locomotives, the speed is measured on the basis of the number of revolutions of the drive shaft and displayed on the speedometer in the driver's cab. With that system, if the drive shaft being used to measure speed happens to slip, it becomes impossible to measure the speed correctly. In order to avoid that problem, a special axle for measuring speed, one not attached to either the electric motor or the brakes, was attached.

In addition to the four types of brakes described earlier, emergency storage batteries were loaded into the locomotives for use in case of a power outage. If a power outage occurs on a steep gradient, it is difficult to descend the gradient on just the air brakes, because of worries that they might burn out.

In the event of a power outage, therefore, the train has to wait until power is restored, but the air in the air brakes is consumed up little by little as time passes. The batteries were fitted to operate the air compressor that replenished the air.

Piling precaution on top of precaution, the designers also equipped the locomotive with a mechanical lever to act as an emergency holding device that pressed the brake shoes against the wheel tread when the train was stopped for a long time.

The locomotive was also equipped with another special device, an overspeed detector. If the train's downhill speed exceeded a certain limit, the emergency brake worked automatically to stop it.

This speed limit was 40 km/h for passenger trains and 25 km/h for freight trains. Above these speeds, the train might be unable to stop.

In fact, during the test runs, I gave the directions to turn off this switch while descending the hill on a freight train. Leaving from the Kumanotaira Signal Station, the test freight train soon started to accelerate as it went down the gradient, and since we didn't apply the brakes, the speed increased as we watched. We suppressed our fear until the speed reached 35 km/h and then applied the emergency brake.

We felt a jolt that threw us forward, and the train gradually slowed, but it leveled off at 25 km/h. It did not stop.

We knew very well that the 500 ton weight of the freight cars was pushing the locomotive from behind. Since we had already applied all the brakes, there was no way of stopping the train. As we approached the Maruyama Signal Station, the gradient became more gentle, and the train stopped on its own, but all in all, the test gave me shivers. Even though I knew we were all right, I worried about what to do if a train happened to run out of control.

The first series of tests that began in June 1962 went relatively smoothly. There were almost no problems with the brakes that we were worried about, but the slipping that occurred when the trains climbed the gradient was worse than expected. We scattered increasing amounts of sand on the tracks to prevent slipping, but in the end, it wasn't enough. We concluded that ordinary methods were insufficient for dealing with a 67‰ gradient.

We discussed a variety of countermeasures, and the second series of tests began in January of the next year. This time, the tests went comparatively well, and we also began training the drivers at the same time. Since it was, in any case, a problematic section, each driver ran about 40 roundtrips during his training. But the trainees didn't just run the train straight through. Their training also consisted of starting up again after stopping on a gradient, procedures for dealing with breakdowns that stopped the train, and so on.

Just as I was thinking that everything had gone well, Jiro Sato, the head of the Yokokawa Engine Depot, came to my office at the Headquarters of JNR. He was a small, slim, earnest man, and his words spilled out in a rapid stammer.

4.5   The Coupler Breaks

> Sir, we have a real problem. The coupler has been breaking sometimes.
>
> Well, that kind of thing does happen from time to time, you know.
>
> No, sir, I just think there's something odd about this.

The coupler had been breaking nearly every day during testing since the beginning of May. This kind of accident had not occurred during the test runs with only one EF63 locomotive as the number of locomotives was small, but had started to happen during the test runs when two EF63s were attached to a train.

When we investigated to see when the mishap occurred, we found that it was when the EF62 was at the front hauling and two EF63s were at the rear pushing a train going uphill. When the safety device known as the overcurrent relay went to work on the EF62 locomotive, the coupler broke. When this relay was working, the front locomotive lost its traction force, so the combined weight of a nearly 100-ton locomotive and 500-ton train was applied momentarily to the two EF63s that were pushing from behind, and the coupler between them broke.

The same type of thing happened when going downhill, with all three locomotives at the front. In this case, when the front locomotive applied its emergency brakes, the coupler broke from the shock of stopping so suddenly.

We therefore had a general idea of what was going on, but it was already only two months before the opening of the new line planned in July. We needed to do some serious testing and put some emergency measures in place. Waiting for construction work on the tracks to be completed, we conducted special tests on the coupler problem on May 16. A number of experts on rolling stock, including Dr. Tetsu Matsui of the Rolling Stock Performance Research Laboratory of the Railway Technical Research Institute, assembled for the occasion.

The test train was an EF62 locomotive in front of a 400-ton freight car (instead of real freight cars, we used decommissioned heating cars) and finally the test cars, with EF63s pushing it from behind. We first tested the case in which the problematic overcurrent relay of the front locomotive was working. Instead of having the overcurrent relay working, we had the driver turn the power notch off. As supervisor of the test, I picked up the microphone.

> Driver, Driver, are you ready? We're going to be starting the overcurrent relay action test. At the signal, turn the notch off. Go ahead.
>
> Yes, this is the driver of the EF62. Preparations are in order. Go ahead.
>
> All right, we're starting the test. Five, four, three, two, one, okay, notch off!

One or two seconds later, we felt a tremendous shock. It felt like an earthquake with an intensity of seven. The test car swayed and seemed about to break up, and I was nearly thrown out of my seat.

Looking at the coupling heads, we found that the freight car ahead of our test car had been suddenly pushed upward and then swayed strongly to the right. Fortunately, the emergency brake was able to stop the train without causing a derailment.

The force of the two EF63s was significant, and they didn't slip, even with this kind of shock. When we hurried to take a look at the measured data about the force applied to the couplers, we found that it exceeded 60 tons. Since the coupler was strong enough to withstand only up to about 50 tons, no wonder it broke. We were afraid to conduct a second test. I rushed back to the headquarters to report our findings and hurried to find the engineers in the Rolling Stock Design Office. However, they didn't trust me at first.

We conducted all the proper tests, and we didn't find any problems, so everything's fine.

That kind of thing doesn't happen very often. It's a freak occurrence.

Mr. Yamanouchi, you've conducted another one of your reckless experiments, haven't you?

I wasn't kidding. Was tripping the overcurrent relay a freak occurrence? It was nothing of the sort. It wouldn't necessarily happen often, but eventually, some abnormal situation would occur, and in order to protect the electrical circuits of the electric locomotive, the protective relay would switch on. Like a household circuit breaker, it would work when it was needed.

An emergency meeting was held immediately, and we conducted tests related to the coupling problem from June 7 to July 2. As we were conducting tests of a variety of scenarios, a derailment occurred. A locomotive was pushing Type 165 EMU train equipped with air springs uphill, and when the emergency brake on the front locomotive was applied, the business-class car in the middle floated up the track and derailed.

This caused a great commotion at the headquarters, and executives came one after another to view the site of the accident.

By this time, I had begun to get an idea of what would occur if I did certain things.

When one of the executives came to the test site, I decided that I would leave him with the strongest possible impression. I had two EF63 locomotives push seven old-style EMUs of the Shonan Type 80 and applied the emergency brake to the main engine, the hindmost EF63 locomotive. The couplers in the middle broke completely. One car came loose and kept rolling, but it soon stopped.

I can still remember the stunned look on the face of the executive watching from the side of the tracks.

As a result of the series of tests, we found out the following:

- During normal operations, the force applied to a coupling was approximately 30 tons, so there was no problem.
- An abnormally high force was applied to the couplers when the overcurrent relay of the electric locomotive was working, when the emergency brake was applied, and when the conductor pulled the conductor's valve to apply the brakes.
- When pushing EMUs, DMUs, and other rolling stock with disk brakes, applying the brakes exerted a huge force, since the brake performances were different.
- Rolling stock with air springs was slightly more likely to derail.
- There was a strong possibility that the coupler would break if the emergency brake were applied in the rear EF63 locomotive when it was climbing a hill.

- We could not operate long EMU and DMU trainsets that required three EF63 locomotives.

July 15, the scheduled opening day for the new line, was coming up fast, but with things as they were, we couldn't switch over to it completely. We therefore decided to postpone the opening of the new line until October and immediately run only semi-express trains using old-style Shonan Type EMU trains on that new line. We did this because old-style Shonan Type EMU trains don't use air springs. Other trains were kept on the Abt system on the old tracks for the time being. As a result, the *Hakucho* limited express and the *Shiga* DMU express had to run on the old tracks, and they were more than 10 minutes slower than the semi-express running on the new line.

Another troubling problem was that since this semi-express was made up of 12 cars, it couldn't climb the hills without the aid of three EF63 locomotives. However, this had been judged dangerous. We therefore divided the train in half at Yokokawa, with six cars headed to Nagano, and six cars going to Naka-Karuizawa 18 minutes later.

We somehow just barely managed to save face, but we would need to have answers by October. After countless meetings, it was decided to insert a throttle valve into the circuit of the conductor's valve in order to make the emergency brake work a bit more slowly, improve the buffers of the locomotives, remove the air from the air springs for this section only, and pull fewer passenger and freight cars than originally planned.

We therefore retested everything several times between August 5 and August 22, and eventually, beginning in October, we got rid of the Abt system and moved all trains over to the new line.

The fact that these kinds of setbacks could occur even after painstaking studies and tests by experts was a tremendous learning experience for us, and the accidents reminded us to be wary of 67‰ gradient. You might even say that the train was able to run safely because we conducted so many tests, and in many respects, I learned a number of valuable lessons from the experience.

Regrettably, however, there was one major accident on this line. On October 28, 1975, four locomotives deadheading from Karuizawa to Yokokawa gained excessive speed on the downhill portion, and applying the brakes did not slow them down. Just after emerging from the first tunnel, they derailed and overturned.

The locomotives were already tilted inside the tunnel, and the walls of the tunnel were still scarred where the locomotives scraped against them.

As I mentioned before, excessive speed on a downhill gradient is truly something to be feared. That's why EF63 electric locomotives were equipped with overspeed detection devices that automatically applied the brakes if the speed exceeded a certain limit. In the case of this accident, however, someone had evidently turned off the power switch for this device for some reason.

This memory-laden section between Karuizawa and Yokokawa was eliminated in 1997, and another Japanese railway famous landmark vanished. In today's world, with the system of expressways expanding, it isn't considered appropriate to seek out sentimental experiences.

4.6   It's the Track that Determines How Fast a Train Can Go

With the appearance of the *Kodama* EMU limited express, trains that could reach 110 km/h were now a reality in Japan, too.

However, Europe and the United States already had express trains that ran at 160 km/h. Admittedly, Japan was handicapped by its meter gauge tracks, but people wondered if better results couldn't be achieved with just a little more effort.

As I mentioned before, regulations already allowed speeds of up to 110 km/h on the main trunk lines. It was just that only the *Kodama* and other EMU limited expresses on the Tokaido Line actually achieved that speed. Why couldn't other trains run faster? What problem was preventing them from running faster, and what could we do about it?

In order to grapple with this problem directly, JNR set up the Train Speed Survey Committee in 1960, and at the first meeting, Chairman of this committee, Shiro Seki, Board member of JNR, spoke as follows:

> Along with advances in the modernization of motive power, there are increasing demands for raising the speeds of express and semi-express trains. The other side of the picture is that the structure of the rolling stock has also been changing, so we must think theoretically and bring about an increase in speed over a broad range without spending much money. I would like to proceed in a two-tier setup. The committee will study the speed increases required by the exigencies of revising the train schedules from scratch and analyze the basic factors related to speed in general. JNR will study the question of how fast trains will go in the future.

This committee's first step in its basic analyses for increasing speed was deciding to study specific areas of concern, and it took up the case of increasing the speed of the *Hatsukari,* which had just gone from being an ordinary passenger train hauled by locomotives to being a DMU limited express. The committee set a target of having the train travel between Ueno and Sapporo in 20 hours, including the time on the ferry over the Tsugaru Strait. This was two hours and 20 minutes faster than a passenger train hauled by a locomotive.

First, the maximum speed would be raised from 95 to 100 km/h. JNR was ready to aim for 110 km/h, but there were limits to the number of revolutions the engine was capable of, and to the braking distance, as well as other problems, so it would be difficult. The regulations had already determined that it was permissible for DMU trains to pass through curves faster than EMU trains and locomotives by five km/h. The remaining problem was strengthening the tracks. The introductory section of the minutes of the first meeting of the Train Speed Survey Committee reported,

> Since raising the speed of a train is directly connected to damage to the tracks, for that reason alone, and now that the tonnage of passing trains has increased, we need to strengthen the tracks.

Then noted,

> Strengthening the tracks in order to increase speeds on one or two specific trains, as with the *Hatsukari* round trip, is necessary, but would it not be possible to apply these increases

in speed to rolling stock and sections that enjoy the same conditions? This is an argument that we have gone over any number of times, but since raising the speed limit has a great influence on damage to the tracks, people have strong opinions to the effect that it is difficult to establish a uniform speed limit. We are painfully aware of the need to investigate such matters as whether this committee will take up the question of increasing the speed on specific trains or whether it will move in the direction of studying track structures and speed limits in general.

In any case, the committee discussed replacing the ballast on the rail beds with crushed rock, placing tie plates between the rails and the sleepers, and other measures for strengthening the tracks, and increased speed for the *Hatsukari* became a reality.

Afterward, the Train Speed Survey Committee turned away from studying these specific areas of concern and began elucidating the fundamental problems related to increasing speed. Besides analyzing such things as the limits of the lateral force applied to a track when a train passes, criteria for riding comfort, and the effects on the tracks when trains run, it also set goals for future increases in speed and made plans for strengthening tracks. It was at just this time that I was put in charge of planning for speed.

The three main players in the arguments about increasing speed were the Train Schedule Division of the Train Operations Bureau, the Permanent Way Maintenance Division of the Infrastructures Bureau, and the Rolling Stock Planning Division of the Rolling Stock and Mechanical Engineering Bureau. We met nearly every week to discuss. As a representative of the Train Schedule Division, I insisted that the new limited expresses such as the *Kodama* were a great success and that almost all the trains were close to capacity. This would be the age of the limited express. "If we are going to bring out the full performance of the EMU and DMU trains," I said, "we ought to increase speeds, and we should be able to."

The Rolling Stock Planning Division also pressed for increased speed. "The new rolling stock for the limited expresses is lightweight, and the suspension devices on the bogie have improved noticeably, so the load on the tracks is far less than with a train hauled by a steam locomotive."

The Permanent Way Maintenance Division was on the defensive and resisted the idea. No, no. We are busy keeping the tracks in good condition for a continuous increase in the number of trains. An increase in speed is unthinkable.

I had thought that the speed of a train depended on the performance of the rolling stock and considerations of safety during high-speed running, but that was not the case. The tracks were the problem. Passing trains wear down the tracks little by little. The question was whether their capacity to compensate for that damage could keep up with the rate of damage. They said that no increases in speed would be authorized on even one line until the tracks were strengthened.

That's strange. The new rolling stock runs incomparably more smoothly than rolling stock hauled by steam locomotives, so there should be less load on the track. The data clearly prove that. At present, the *Hatsukari* and other DMU limited expresses are taking curves at five km/h faster than other trains. What's wrong with EMU trains? Isn't something odd going on here?

The Permanent Way Maintenance Division also had a counter-argument.

The reason that special higher speeds were authorized for DMU trains is that originally, they were just one-car or two-car trains running on local lines. They were light and didn't place much of a burden on the tracks. Yet who would have thought that this regulation would come to be applied to a limited express with thirteen cars and two engines per car? You didn't consult with us or say one word to us, on the grounds that these trains had been authorized by the regulations. The people at the work sites are upset. We have no idea what you people will do once something is authorized.

Well, is there something wrong with that? We ought to be able to do it within the range of what the damage to the tracks will allow.

That may be true, but we would like to put limits on the number of trains that can be speeded up. We'd like you to consult with us each time you increase the number of runs.

You're kidding. Do you think we can do that for every single one? First of all, there is no way that the speeds of the morning and afternoon trains are different.

The vigorous exchange continued for a while.

In the end, it was decided that the type of rolling stock, not the number of trains that run, would be restricted. However, the track people wanted to impose limits so that speeds for certain types of rolling stock types could be increased only if they were used for limited expresses. They wanted to keep the number of high-speed trains low, because it was supposed to be the limited express which need speed up first and foremost.

Now it was my turn to make a counter-argument. "Aren't we supposed to be discussing technical problems? Even though we ought to determine speeds based on the weight of the rolling stock and the characteristics of the run, it's strange that we seem to be making distinctions on the basis of business concepts, such as a limited express or an ordinary express. Are you saying that if you use the same rolling stock, an ordinary express will damage the tracks more than a limited express? If that's the case, let's make everything a limited express."

I probably spoke a little too strongly, and the argument became even more complicated.

4.7   The New Concept of 'Superior High-Performance Trains'

As a result, we gave up the designation 'limited express' and created the new concept of 'superior high-performance trains.' Special speeds would be authorized only for this new class of trains. Regulations for train speeds are still based on concepts from that time. Incidentally, the superior high-performance trains were defined as 'a limited express train or ordinary express train with a limited seating capacity made up of EMUs or DMUs having a bogie with good running characteristics, the axle weight of which is 13 tons or less with a full passenger load, the load under the springs of which is two tons or less, and which uses air springs.'

The reason the definition was worded this way is that the Permanent Way Maintenance Division accepted special speeds for limited expresses because they have, in principle, a limited seating capacity, so it was expected that ordinary expresses with a limited seating capacity would also be acceptable. You probably think that this is a very bureaucratic and opaque way of expression, but it was really difficult for us to arrive at this wording.

Somehow or other, we were able to create rules for increasing speeds. Next, we needed to set goals for increasing speeds in the future. Since this was a significant dream of ours, it was an enjoyable task. Even the Permanent Way Maintenance Division was willing to raise the speed to 120 km/h, on the condition that the tracks would be strengthened against future problems, and they also agreed to raise the speed at which the trains could round curves to five km/h higher than the speed of DMU limited expresses. Based on these assumptions, we made specific plans for speeding up the trains on the main sections and for strengthening the tracks.

The investment in facilities needed to achieve this increase in speed would amount to 210 billion yen over 15 years. Speeds of 120 km/h would first be achieved by the *Hibari* limited express between Ueno and Sendai and the *Toki* limited express from Ueno to Niigata with the schedule revisions of October 1968.

You are probably wondering why we didn't aim to raise the speeds above 130 km/h, but there were several problems involved.

First of all, with the exception of the *Shinkansen*, trains on Japanese railways are supposed to be able to stop within 600 m after applying the emergency brake, in principle. At speeds exceeding 120 km/h, it would be very difficult to abide by this principle.

Next, Japanese railways, with their many curves and hilly sections, have fewer sections than you might think where it is possible to reach the maximum speed. Beginning in 1989, JR East's *Super Hitachi* began operating between Ueno and Mito on the Joban Line at speeds of 130 km/h, but actually the sections where the trains can achieve the maximum speed amount to no more than 28% of the route. Besides, an urban commuter express is often running in front of the *Super Hitachi*, so this further cuts down the sections where it can achieve 130 km/h.

The Chuo Line running through the mountains between Shinjuku and Matsumoto can run at its maximum speed of 130 km/h for no more than six% of its route. Speeding up the trains on Japan's meter gauge tracks to more than 130 km/h is not only technically difficult, but would also require a huge expenditure of funds to achieve a relatively small effect.

In any case, the basic rules established for train speeds at that time were the deciding factor in making Japan the 'Kingdom of EMU Trains'. That's because high-speed trains must be either EMU or DMU. Japan differs from Europe and the United States decisively on this point. Its status as a late-bloomer in building railways, its meter gauge, fragile tracks, and the relative lack of tracks at its major terminals have all combined to make Japan the Kingdom of EMU Trains and the Kingdom of DMU Trains.

To be honest, I was quite at a loss about the stubborn resistance of the Permanent Way Maintenance Division during the speed tests. No matter what I said, they didn't like it. They just kept saying, "If you're going to speed up the trains, then strengthen the tracks."

I even tried badgering them: "Are you bastards going to hold the speed program hostage to get yourselves a budget for strengthening the tracks?"

The tests were almost over, and we were about to decide on the new rules. At dinner one night, when everyone had been drinking, the site representative from the Permanent Way Maintenance Division suddenly blurted out, "You must think we're a bunch of stubborn idiots. But look at it from our point of view. We really envy all of you. You're going to run new trains. They're going to build new rolling stock for that. You just say that you're going to double the tracks and electrify them, and everyone agrees right away. But then we say that you need to strengthen the tracks, and it's, 'We don't have that kind of money,' and that's the end of it. The number of trains has been increasing all along. If you don't strengthen the tracks now, you'll have trouble later. That's why we have no choice but to resist speeding up the trains."

His words left an impression on me. We had some pretty violent arguments, but in the end, I benefited from them, because I came to know about the problems of the other departments and the realities that they faced. I was made to learn about things like the tonnage of passing trains and the relationship between speed and damage to the tracks.

4.8   The Ordeal of Track Maintenance

Around this time, the Permanent Way Maintenance Division, which was responsible for repairing tracks, was also faced with a period of major reforms. You might say that for a long time, JNR's tracks had been maintained by the craftsmanship and loyalty of the frontline employees. In the old days, there was an organization called a track maintenance gang every five or ten kilometers, and they had the responsibility for their own tracks. Ichizo Horikoshi, formerly a specialist with the Railway Technical Research Institute, described the situation as follows:

> No matter what, the track maintenance employees gave their utmost loyalty to JNR, and they made their own section of track a complete work. If I would put it in extreme terms, they finished the tracks as beautifully and neatly as if they were making a miniature garden. They used railbed ballast—the small pebbles that are spread under the tracks—but they finished off the ridge of earth under the railbed ballast so that it was perfectly horizontal, which shows you how much enthusiasm for his work that the work crew manager, the top-ranked leader on the site, had. Once a year, they held a major track inspection, and it was quite an event. The object was to be graded as superior, and sometimes they'd work until midnight, carrying a lantern as they inspected their own section of track. Then they'd make plans for the next day. Some of the men even brought their wives along and had them work alongside them. That's the kind of serious efforts they put forth. People today might see it all as a feudal attitude, but that was the way everyone thought in those days, and the people in charge really relied on them.

Horikoshi said all this in 1973 while reminiscing about the old days. He went on to say,

But the world's ways of thinking changed, and the labor force also decreased. Trains began going faster, and it became necessary to incorporate other systems into track maintenance. There were already thoughts of modernizing the work system with strengthening track structure, that is, increasing the strength of the structures as much as possible and reducing human labor, in the late 1920s. In the end, the question of how to rebuild the tracks that had been destroyed or damaged during the war became a huge problem. There were a lot of devices and developments, and things are still like that today. It was a tremendous change.

What kind of changes happened to the tracks? They may be broadly divided into changes in the structure of the tracks and changes in repair methods, or, in a word, creating sturdy tracks that many trains can run on at high speed, and changing from the former system of repairing the tracks little by little by hand every day to a method of planned repairs using machines.

Ikuma Hiroi, the man I so often argued with during that time, afterward recalled this period.

The former era of depending on superhuman feats of track repair by the work crew manager is over. No matter what, the manager has to stand at the forefront and manage repairs according to a plan. So we also have to change from the method of on-the-spot repairs to the method of regular, periodic repairs. It was during the initial planning for track maintenance modernization that the most groundbreaking reforms in the ninety-year history of track maintenance were made.

He continued,

These plans for modernization were agreed to at the board meetings, but when it came to the implementation stage, we couldn't get you to understand anything about strengthening the tracks. I mean, you had to make huge investments to enhance transport capabilities, and somehow, you never got around to including strengthening of the tracks in the budget. Nearly every time, Kubota or Yamanouchi said, 'Won't you let us raise the speed on just one limited express?' but there are two reasons that we didn't just nod and agree. The first is that we didn't have the confidence at that time that these sorts of speeds could be maintained on even one train. Of course, if the tracks were strengthened, it would have been a different story, but as the amount transported increased, on-site maintenance became increasingly difficult. The second reason is that even if the tracks were improved to a certain extent, we wondered what would guarantee keeping to one train. I think we got everyone's agreement there should be some limit on the increase in the amount of damage by combinations of a fleet of high-speed trains and the kinds of high-performance cars that are suitable for speeding up.

The tracks of Japanese railways have several handicaps, compared to those in Europe or America. The first is that meter gauge tracks have poor resistance to the weight and impact shock of having the trains run on them. Given Japan's volcanic geology and the alluvial soil along rivers, the ground is weak in many places, and there's also a lot of rainfall. The tracks bear a heavy load because of many curves and gradients. Since there aren't numerous tracks on the main lines or in stations, a lot of trains run on each track. Therefore, it's hard to adopt such methods as stopping the trains by adopting the single-track operations temporarily so that we can carry out intensive track repairs. So what was to be done about strengthening the tracks? First of all, the rails could be made thicker. The rails used at that time on trunk lines weighed about 50 kg per meter, and even thinner rails were used on some sections.

We would increase the number of sections where heavier standard rails were used. Recently, we have been using even heavier rails of 60 kg per meter.

The most radical technology for rails was the adoption of long welded rails. Formerly, the length of a standard rail was 25 m, and they were held together with fishplates. Rails expand and contract with changes in temperature, and the length of a 25 m rail changed about 20 mm between summer and winter. It used to be conventional knowledge that rails would warp in the summer if you didn't leave gaps at the joints in order to absorb the expansion and contraction.

4.9 Long Rails and PC Sleepers

However, this conventional knowledge was first overturned in Europe in the 1920s. There was research into the question of whether rails would be unaffected by changes in temperature, even if the rails were welded and long, if they were adequately supported by sleepers.

Tests of long rails began in Japan, too, in 1927, and in 1934, long rails of 250 m were installed experimentally on the Tokaido Line.

At about the same time, Dr. Ichizo Horikoshi of the Railway Technical Research Institute independently announced a theory on long rails. He installed five km-long rails inside one of the tunnels on the Senzan Line and began collecting actual data.

It was after World War II that Japan began to get serious about laying long rails. This type of rail began to spread gradually in or around 1953, along with advances in welding technology.

Long rails were adopted for the entire length of the Tokaido *Shinkansen*. Even so, the distance between Tokyo and Osaka can't be covered by just one rail section, and since the rails need to have insulated parts for the signal circuits, the rails have been made to be about 1.5 km in length.

Long rails are an essential technology for the *Shinkansen*. One of the biggest problems in creating the *Shinkansen* was whether it would be possible to repair the tracks properly so that they wouldn't be harmed by high speeds.

Facilities for high-speed running must be prepared with greater precision than facilities for conventional lines. This is because basic safety depends on it. On the other hand, fast trains place a greater burden on the tracks than do slow trains. Especially if rails have joints, failure of that part can warp the rail. An extremely important challenge for the *Shinkansen* was getting rid of the weak point of the track, the joint.

Reading the report of the *Shinkansen* Construction Standards Fact-Finding Committee gives one a feeling for the importance of the problem of track repairs. A track expert said,

> There's no problem with an EMU train for commuters, but we aren't confident of our ability to repair tracks for high-speed running such as the *Shinkansen*. Some adjustments will be needed during the day, but there won't be enough of an interval between trains to allow that, so there will be constant slowdowns.

Committee Chair Hideo Shima countered,

Then we'll have to make it possible for you to repair the tracks at night. Not being able to make repairs means that the trains won't be able to run for three hours.

Then the Permanent Way Maintenance Division offered a plan in which turnouts would be built along the track and the trains would run partly on a single-track basis so that repairs could be made during the day.
The response to this was,

Do you mean putting in turnouts every 10 kilometers?
   We would need about 50 of them along the entire line.
   So then every day the trains will be running on a single track somewhere?
   For the most part, it will be one place per day.

A committee member from the Train Operations Division replied,

All we have to do is to lay three tracks when we refurbish them, and then the trains can run slowly on two tracks. I think that running on just one track will lead to gridlock. It's better to have the trains run on double tracks than on single track, even if they have to go slowly.

The Permanent Way Maintenance Division insisted that it was quite impossible to perform repairs only at night. Unless three tracks were laid or turnouts were set up at intervals with the trains running on single tracks during daytime repairs, they would not be able to repair the tracks.
But a committee member responsible for the meeting coordination mentioned,

A three-track plan and a single-track plan have been proposed. The three-track plan would increase construction costs, and there would be both technical and practical problems with single-track operations, so it's a perplexing situation.

In the end, JNR set aside the period from midnight to 6:00 a.m. as the 'work period' for the *Shinkansen*, a time when no trains would run and repair operations could be carried out. It also decided on special procedures. This may have been the first case of its kind in the world, and even though it's an unexciting system, one can say that it is the basis of the *Shinkansen*'s safe operation (Fig. 4.3).

About half the length of track in the JR East system is currently of long-rail structure. In addition, recent developments in signal technology make it unnecessary to install joints, and the *Shinkansen* between Takasaki and Karuizawa uses 40 km-long rails without joints.

When the rails are welded together and made longer, they naturally expand in the summer and contract in the winter. Yet one cannot make joints with large gaps. It is therefore necessary to hold the rail down securely with sleepers in the middle so that they neither expand nor contract.

In addition, since great force is applied to the sleepers, they need to be sturdy and heavy. Existing wooden sleepers lack the necessary strength, and heavy concrete sleepers are more suitable. For the ballast, too, we need to lay down angular crushed rock instead of round pebbles.

Fig. 4.3 Work crew repairing the tracks with the multiple-tie tamper maintenance car after the last *Shinkansen* of the day has passed through. *Photo* Provided by Kotsu Kensetsu

Since concrete is extremely sturdy against compressive forces, it is appropriate for such purposes as supporting heavy objects, but it is unexpectedly weak against tensile force. A train passing over concrete sleepers exerts a tremendous tensile force on their upper surface, so if nothing is done, cracks develop. Evidently, when research was conducted on concrete sleepers in Japan before World War II, the researchers had trouble with cracks in the sleepers due to this very problem.

Therefore, if steel rods are embedded in the concrete when the sleepers are manufactured, the rods perform the role of withstanding tensile force, so the concrete becomes difficult to break. This is the ferroconcrete principle used in building construction, which allows both the steel rods and the concrete to make the best of each other's strengths and compensate for each other's disadvantages.

Furthermore, if the steel rods are subjected to a strong tensile force before the concrete hardens, the result is a ferroconcrete component that is even more resistant to tensile force. The sleepers produced according to this technology are referred to as 'PC concrete ties,' and the technology allows the use of concrete in components that undergo violent impact shock, as railway sleepers do.

Ferroconcrete is said to have been used first in France in 1849 as a shipbuilding material. The French engineer Joseph Monier took out a patent on ferroconcrete railway sleepers in 1877 and also built ferroconcrete bridges. This was at a time just after the first railways went into operation in Japan.

Research into PC concrete technology began in about 1888, but it was only in the 1920s that it became a really practical technology, and another French engineer Eugène Freyssinet was the pioneer in this area. Even if one understood the theory of making ferroconcrete, it seemed difficult to come up with practical applications until Freyssinet succeeded in devising a manufacturing method that would hold the steel rods neatly in place in the concrete.

JNR learned this technology and others, from Europe, tested it repeatedly, and determined the standard design, using a German design for railway sleepers as a sample. PC concrete ties were first put into actual service on such main lines as the Tokaido and Sanyo around 1950. The Tokaido *Shinkansen* required 1.5 million of these sleepers, and their standard design was determined on the basis of comparison between a variety of types. It then took two years to manufacture them.

When wooden sleepers were replaced by concrete sleepers, the fasteners that held the rails to the sleepers also had to be changed. Wooden sleepers were fastened with dog spikes, but a new method had to be created for concrete sleepers.

Japan adopted a method based on a French one, in which springs were positioned in above the sleepers and were fastened with bolts. Two types of springs served to decrease impact shock and vibration when a train passed over. One was a small plate spring placed between the bolt and the rail to soften vertical and horizontal vibration. The other was an iron plate, called a 'tie plate,' laid between the rail and the sleeper. In addition, a thin rubber sheet was laid between the rail and the tie plate.

This fastener has superior characteristics, and it became the standard form for JNR, especially on the Tokaido *Shinkansen*, but it was sometimes necessary to inspect and tighten the bolts. After its launch, JR East tested a fastener using a small coil spring from the British company Pandrol, and we found that it reduced the need for maintenance, so we have been using it widely. If you stand on the *Shinkansen* platform at Tokyo Station and look at the tracks, you ought to be able to see these fasteners. In the world of railwaying, each country has its own technology, which you can see if you take a look at the fasteners holding the tracks together.

Even long rails need joints. Since the concrete sleepers and the fasteners are firmly holding the part between the rails, the length of the rail does not change, even if the temperature changes or stress accumulates. However, either end of the rail may expand or contract somewhat due to temperature variations. The earlier types of rails were simply held in place by the bolts securing the fishplates, but there was a great deal of impact shock when trains passed over a joint part at high speed. This not only made the ride less comfortable but also caused severe damage to the rails.

Expansion joints (EJ) were the result of creative thinking about a mechanism that had no gaps and could smoothly absorb the expansion and contraction of the rails. After comparing the expansion joints used in various European countries, Japan decided to model its expansion joints after a French type (Fig. 4.4).

French technology has had a great influence on the structure of Japan's new tracks. The reasons are France's pioneering role in structural engineering, exemplified by ferroconcrete and the Eiffel Tower, and its experience with operating high-speed rail

Fig. 4.4 Synthetic sleepers and the expansion joints for long rails. *Photo* Provided by JRTT

lines including its high-speed tests of up to 331km/h. Yet there is also the fact that France also welcomed Japanese engineers to study in their country. Many young specialists in railway technology took advantage of this system of study abroad in order to go to Europe. These people formed a group called 'SABTECH,' and they are still continuing these friendships.

At a lecture meeting of the Railway Technical Research Institute in 1957, Dr. Yoichi Hoshino, an expert on railway tracks, suggested that the new tracks for high-speed railways not be built with ballast laid underneath. Instead, he proposed attaching the rails directly on top of concrete slabs with some sort of buffer material in between. Although it was not possible to apply this new track structure on the Tokaido *Shinkansen*, it did replace the older structure beginning with the Sanyo *Shinkansen*.

4.10 The Battle with Sparks

Even though SNCF was able to achieve a record speed of 331 km/h in its high-speed tests in 1955, the pantograph of the test train was burned off due to arcing, and the rails that the train had passed over were softened and deformed.

This was clearly unacceptable for commercial operations. In fact, in the nearly 50 years since that test run, no train has ever run at 330 km/h.[2] Success in a test

---

[2] Currently, several lines in China, including the Beijing-Tianjin route, operate at a maximum running speed of 350 km/h. In France, TGV-POS has started 320 km/h operation on the Eastern Europe Line, which opened in 2007.

run and success in commercial operations are entirely different, and sometimes a lot of problems remain between a flashy, much-talked-about topic of successful commissioning and commercial operations. One of the difficult problems we had to solve in order to make the *Shinkansen* a reality was that of arcing on the pantograph.

The power source for an EMU train is the overhead contact wire strung over the tracks, which supplies electricity to electric motors through the pantograph. If you look at a *Shinkansen* train running at night, you can see that it constantly gives off little sparks above the roof. This amount of arcing is no problem, but if the pantograph somehow moves away from the overhead contact wire, the result is large sparks that can melt the pantograph or break the contact wire.

At high speeds, it becomes very difficult to collect current through the pantograph. It can be blown off by a strong wind, and the overhead contact wire may begin to vibrate a bit. The amount of current flowing increases greatly. For this reason, creating an overhead contact wire and pantograph that would keep continuous, smooth contact and not give off huge sparks was one of the biggest challenges to realizing the goal of high-speed running.

The pantograph on a *Shinkansen* train is much smaller than that on an EMU train for conventional lines, and its shape is somewhat unusual. A pantograph appropriate for high-speed running is as small as possible with little inertia. A strong spring is less likely to allow the pantograph to separate from the overhead contact wire, but if the spring is too strong, it may damage the contact wire.

The wire that conducts the electricity is a contact wire which is suspended from above by another wire called a 'messenger wire' and is in contact with the pantograph. This double structure, referred to as 'catenary,' makes the overhead contact wire uniformly flexible so that it is less likely to separate from the pantograph. Despite such contrivances, some vibration occurs when the pantograph runs to pass the points where the contact wires are supported, so the pantograph separates from the overhead contact wire, making arcing likely.

Hoping to prevent this phenomenon somehow, the Railway Technical Research Institute proposed an overhead contact wire with a special structure called a 'continuous mesh-patterned overhead contact wire,' and they tested it, but it was not adopted because its structure was too complicated.

A composite compound overhead contact wire was adopted for the Tokaido *Shinkansen*, with three catenary wires and a small damper for preventing vibration inserted on the vertical wires from which the contact wire hangs. The structure is rather complicated, but the designers have succeeded in creating a wiring system that produces few sparks (Fig. 4.5).

France's TGV, which was completed 17 years later than the Tokaido *Shinkansen*, mounted a frontal challenge to Japan's technology. The pantograph was large instead of small, but it lacked the half-diamond-shaped part. SNCF had been using this unusual pantograph for a long time on electric locomotives, and it is lightweight, with low inertia, although the wind pressure on the pantograph varies slightly depending on the direction in which the train is running.

Fig. 4.5 Simple modern pantograph. (Pantograph of the E6, successor to the E3 series for the *Komachi*.) *Photo* Provided by JR East

JR East installed one of these French-style pantographs on its experimental EMU train the *STAR21* and compared it with Japanese pantographs, but we didn't get the impression that its current collection performance was particularly good. That may be because it didn't quite match the performance of the Japanese overhead contact wires. Afterward, we made improvements to the 'single arm pantograph' and are using it on the E3 railcars for the Akita *Shinkansen* (Fig. 4.5).

The complicated wiring system devised in Japan was not adopted for France's TGV. Instead, the French adopted the simple structure used for trains on the conventional lines, the 'simple catenary suspension,' and strengthened the tensile force in the overhead contact wire. That may have been a challenge to Japan's *Shinkansen*, and in fact, the *Shinkansen* wiring has recently been changed to a simpler structure. As we can see just from the example of current collection technology, the sparks of technological innovation have been flying back and forth between Japan and France.

4.11 From Military Researcher to Railway Researcher

Philippe Roumeguère, former vice-president of technology and development for the French National Railway (SNCF) and Director General of the International Railway Union (UIC), once shared his thoughts with me.

Japan's construction of the *Shinkansen* was quite a stimulus to us. That was one of the reasons I chose to go to Japan on my honeymoon. During our stay in Japan, I spoke about the *Shinkansen* with a number of people and read a lot of materials, but my strongest impression was of the leadership provided by Hideo Shima and the Railway Technical Research Institute.

> I was very impressed by the fact that people who had been in military research institutes during World War II moved to the Railway Technical Research Institute after the war, and that their theories and research contributed greatly to making the *Shinkansen* a reality.

As a matter of fact, the role that the Railway Technical Research Institute played was indeed a significant one. You might say that the three entities joined together in this process of bringing the *Shinkansen*'s technology into being: JNR Headquarters, with its planning ability and expertise and basic design technology; the industrial sector that supports the railways with its technological developments, and design / manufacturing technology; and the Railway Technical Research Institute, with its competence for analysis and development technology in basic areas.

The Railway Technical Research Institute was deeply involved in the design of the *Shinkansen* from the beginning. In May 1957, it held a lecture meeting to commemorate the fiftieth anniversary of its founding, and the theme of the meeting was 'Super Limited Express Trains: the Possibility of Traveling between Tokyo and Osaka in Three Hours.' These were the first specific technological proposals for the concept of the *Shinkansen*, and they were a major first step in making the high-speed system a reality.

Takeshi Shinohara, who was Head of the Railway Technical Research Institute in those days, told about this lecture to me as follows,

> France's test run of 331 km/h was what provided a stimulus for us back then. I thought it was odd that in this day and age, Japan's finest scholars could get together and do research and end up not being able to do anything. That's why we talked among ourselves and decided to discuss the possibility of traveling between Tokyo and Osaka in three hours, so we got permission from President Sogo and Chief Engineer Shima.

There were four presenters at the lecture meeting: Tadanao Miki, Head of the Rolling Stock Structural Research Laboratory, spoke about rolling stock; Yoichi Hoshino, Head of the Track Research Laboratory, spoke about tracks; Tadashi Matsudaira, Head of the Rolling Stock Dynamics Laboratory, speaking about riding comfort and safety; and Hajime Kawanabe, Head of the Signal Research Laboratory, speaking about signals and safety.

Of the four speakers, three of them, excluding only Hoshino, who spoke about tracks, a field specific to railways, were researchers who had formerly worked at military research institutes and had been involved in the research and development of airplanes, ships, weapons, and similar items. After the war, they moved to the Railway Technical Research Institute to find a new sphere of activity. The story was the same for nearly a thousand former military researchers who had transferred to researching railways.

During the war years, most of the very best science students joined the military research institutes, and they were dealing with the most advanced research in many fields. The situation was the same in the United States and Europe, and the Apollo program, the space shuttle, and the Internet are only a few examples of advanced technologies that the armed forces were involved.

These researchers in completely unrelated fields who came in from the military were like a breath of fresh air in technological research for railways.

Dr. Miki worked on making rolling stock more lightweight and on problems of air resistance. Dr. Matsudaira, Head of the Rolling Stock Vibration Laboratory, was an expert on vibration problems, and in naval research organizations, he had been involved in researching problems of vibration in aircraft.

When aircraft reach a certain speed, they sometimes start to vibrate violently, and in the worst case, they may break up in mid-air and crash. This phenomenon is called 'flutter.'

This is the problem that Dr. Matsudaira specialized in. It was the phenomenon called resonance or self-excited vibration in vibration theory, the most basic problem in the field and known by anyone who has studied it even a little. Yet it is not as simple as the theory would suggest to calculate the speed (or more accurately, the frequency that will cause self-excited vibration) in an actual aircraft or to come up with a design that will prevent flutter from occurring in an actual flight.

Dr. Matsudaira analyzed a great deal of data and developed a theory for calculating the specific frequency that would cause flutter.

The same kind of phenomenon occurs in rolling stock. At a certain speed, the railcar begins vibrating in all directions, leading to 'hunting,' a phenomenon in which the railcar sways laterally as it moves forward. If a train is hunting particularly vigorously, one has to worry about a possible derailment. This was the biggest problem related to running the *Shinkansen* safely at high speeds.

Of course, the Railway Technical Research Institute had been studying the problem of vibration in rolling stock for a long time, and it had specialists on the staff, but most of their knowledge had been gained through experience, and evidently, no theoretical analysis had been performed.

Dr. Matsudaira recalled the first time he visited the Railway Technical Research Institute.

I was surprised when I stepped inside the gate. It was just three or four plain wooden barracks, and I couldn't see anything that looked like a real research facility. 'Oh, no,' I thought, 'is this the Research Institute?' I had grown accustomed to the Naval Aviation Technical Arsenal, which in those days was considered one of the foremost, most splendid research institutes in the world, and no matter how I looked at it, I just couldn't believe that it was a research institute.

Then he met a specialist in rolling stock vibration at the Institute.

He seemed like an extremely gentle-mannered person, and when I said that I wanted to research problems of vibration in rolling stock, he told me, 'So read this.' and handed me a bunch of papers and reports that he himself had written, there were more than a dozen of copies. I went home thinking that there was nothing left for me to do if he had done that much research. But once I read the papers at home, well, it sounds awful to say this, but I was really relieved. I mean, they were called scientific papers, but they were just compilations of the results of measuring vibration in rolling stock up till then, and they just described whether a certain car experienced a lot of vibration or very little. There were some places with a little bit of theoretical treatment of the subject, but the theory was rudimentary, and there were no attempts at a thorough theoretical analysis of vibration in rolling stock. Honestly, my first impression was that there were any number of things left for me to do.

How did the ride on JNR's rolling stock seem to a vibration expert like Dr. Matsu-daira, right after the war, when he had just moved to the Railway Technical Research Institute? According to him, "The vibration in the rolling stock at that time was several times worse than it is now, and the vibration in the EMUs was striking. Riding comfort was extremely bad. I never thought that these kinds of EMU trains could be used as long-distance trains."

It's a situation we can hardly imagine, living in an era when almost all limited express trains, including the *Shinkansen*, are either EMU or DMU.

4.12   Conquering Vibration with New Bogies

Perhaps stimulated by the concepts brought in by these kinds of new researchers, JNR formed the High-Speed Bogie Vibration Study Group around Hideo Shima, Head of the Motive Power Division of the Rolling Stock and Mechanical Engineering Bureau in December 1946.

This study group was the center of research for the rolling stock design engineering corps of JNR Headquarter, the researchers of the Railway Technical Research Insti-tute, and specialists from the manufacturers of rolling stock. You might call it the take-off point for Japan's postwar technological development of high-speed railways.

As you remember, 1946 was only a year after the end of World War II, and the country was in a state of postwar confusion, and yet this kind of research was already underway.

Furthermore, in March of that same year, electrification of the Joetsu Line between Takasaki and Nagaoka began, and enthusiasm for this kind of progress in railway technology eventually led to the creation of the Tokaido *Shinkansen*.

This study group evidently met six times in three years, but as a result, the ride and running performance of EMU trains, from the Shonan EMU trains to the *Kodama* limited express, improved remarkably. The former taboo that 'it is difficult using EMU trains for limited expresses because their ride was so bad' disappeared. There is no doubt that these kinds of technological developments created Japan, the Kingdom of EMU Trains, and made the *Shinkansen* possible.

This research was bearing fruit right at the time I entered the engineering faculty of my university, and bogies with new designs were appearing on the scene one after another. That was one of the reasons I chose the bogie of an EMU as the theme of the design for my graduation thesis. I also read Dr. Matsudaira's papers and studied the new types of bogies of various designs in Japan and overseas.

When designing a bogie, you have to consider the problem of vibration and the problem of strength, and I suppose you can say that's enough. I wanted to decrease vibration so that the train could run safely, but what could I do about a smooth ride? At the same time, could I minimize the impact shock to the track? In order to do that, I mounted the heavy components, such as the electric motor, on springs as much as possible. The position that transmits the traction power of the bogie to the frame had to be as low as possible. I gave up using crude leaf springs, which were likely to cause vibration, and replaced them with more flexible coil springs. Choosing the degree of flexibility of the springs was an important design consideration, and here Dr. Matsudaira's paper provided the basic knowledge that I needed. I determined the combination of spring and hydraulic damper (a vibration control device) that would not produce abnormal vibration, similar to flutter in an aircraft.

Theoretical analysis of the old bogies with leaf springs was quite difficult, but the combination of the coil spring and the hydraulic damper resulted in the theoretical analysis and the actual vibration matching beautifully. As to the question of how much vibration would make a passenger feel uncomfortable, there was Janeway's research on ride criteria, which presented the limits of tolerance for changing vibration based on the frequency of that vibration.

EMUs vibrate not only vertically and laterally but also in motions such as rolling and pitching. In fact, there are six kinds in all. The most annoying problem is the type of vibration called hunting, and it not only makes the ride unpleasant but can also lead to derailments if it becomes really severe.

In order to prevent hunting when the bogie begins vibrating abnormally, you can make sure that friction occurs between the carbody and the bogie to reduce vibration, or you can use a hydraulic damper, but it is important to attach the axles firmly to the bogie because if these parts are at all loose, it can lead to hunting.

Even so, since there are springs for absorbing vibration between the bogie and the axle bearing, a certain amount of vertical motion is necessary. Determining the structure of this part is another point to consider when designing a bogie.

Europe was more advanced than Japan when it came to researching and developing new types of bogies. Even the High-Speed Bogie Vibration Study Group seems to have begun its deliberations by surveying and analyzing new bogies from overseas. There was Swiss company Schlieren, whose method required the axle bearings to be firmly in place while providing flexible support. The French company ALSTOM had created its own unique design, and DB's Minden Research Institute had also decided on a unique design and was already using it in its standard passenger cars. Japan's manufacturers of rolling stock either acquired licenses for these new European design technologies for bogies or announced their own unique designs.

In order to determine the design of the bogie of the *Shinkansen*, JNR set up the *Shinkansen* EMU Bogie Study Group in July 1958 and called for designs of new high-speed bogies from Japan's manufacturers of rolling stock.

Proposals came in from six companies, and most of them were based on new European designs. When JNR tested these six kinds of bogies on its high-speed test platform, none of them were clearly superior or inferior in running performance.

It was finally decided to adopt a uniquely Japanese design for the bogie on the basis of the flexibility of its springs and its ease of maintenance. However, the support method for the axle of this new bogie closely resembled that of the Minden Deutz bogie (used since 1952), the standard bogie for the German National Railway (DB), and one couldn't deny that it seemed like an improved version of the German model.

In addition, the 'WN Driving System,' a system that Westinghouse and Natal developed for the New York City subway, was adopted. It was also adopted for use on the Marunouchi subway line in Tokyo.

About this time, the Railway Technical Research Institute, which Dr. Matsudaira had disparaged as 'barracks,' was rebuilt in the Tokyo suburb of Kunitachi as a modern research facility occupying 220,000 m^2. The rebuilding project was made possible by President Sogo's enthusiasm for the *Shinkansen*.

Chapter 5
Computers Appear on the Scene

This chapter first introduces the hardware that makes up the *Shinkansen*, such as rolling stock, civil structures, and electricity supply, as well as the information systems and software that are indispensable for the safe, high-speed operation of many trains, such as ATC, MARS, and patterned scheduling.

Next, the author will introduce how initial troubles were dealt with in order to enhance safety. No matter how many tests a new system undergoes, some issues cannot be identified until it is actually operated. The accumulation of efforts to find various troubles that occurred after the opening of the *Shinkansen* was important in creating a safe *Shinkansen*. For this reason, the author dares to introduce problems that are not usually disclosed and talks about how the new system, made possible through the accumulation of individual technological innovations, is being polished to the last detail.

5.1 The Development of ATC[1]: The Deciding Factor in Safety

In addition to a derailment, the most frightening kind of accident for a *Shinkansen* train running at high speeds would be a collision with another train. If such a collision were to occur, it could be inconceivably catastrophic. Just designing a high-speed bogie would not guarantee the safety of the *Shinkansen*, not unless we thought up a system to prevent these kinds of collisions.

Given the question of which safety devices would prevent trains from colliding, JNR took the 1962 Mikawashima accident as the impetus for moving ahead with the installation of ATS (Automatic Train Stop) devices on all lines. Yet the ATS systems

---

[1] Automatic train control (ATC) used in Japan's *Shinkansen* is a continuous control safety system based on track circuits. It is equivalent to Level Two of the European Train Control System (ETCS).

© Japan Railway Technical Service 2024
S. Yamanouchi, *If there were no Shinkansen*,
https://doi.org/10.1007/978-981-99-8890-7_5

on the conventional lines did not necessarily function at a level high enough to serve as safety devices for the high-speed *Shinkansen*, so we could not adopt them as such. An even greater consideration was that the very act of operating a *Shinkansen* train at more than 200 km/h while observing ground signals would not be easy. The minutes of the *Shinkansen* Construction Standards Survey Committee describe the problems as follows:

> The maximum braking distance for a *Shinkansen* coming to a complete stop from its maximum speed is about 1,000 m on a level section. The distance at which a ground signal can be perceived is about 800 m, and we believe that it would be difficult to improve on that. When running at a speed of 200 km/h, the time in which one can identify a ground signal is no more than about 14 seconds on a straight section, and since we cannot guarantee an unobstructed view of 800 m on a curved section, the time is further constricted. Since it is impossible to guarantee an unobstructed viewing distance corresponding to the running speed, we can say that a system that requires the driver on board to identify the instructions from a ground signal from inside the train while running at high speeds is unreliable. Therefore, we should install cab signals of highly reliable onboard security mode on the *Shinkansen* so that the current signal can be displayed continuously inside the train. In this case, it would be easy to add the cab signal to an ATC (automatic train control) device.

The basics for the ground facilities, including the structures and standards for the tracks and electrical facilities, were studied and determined by the *Shinkansen* Construction Standards Survey Committee. It was at its 19th meeting in May 1961 that the committee took up the issue of the signaling system.

According to the records from that meeting,

> Since the planned maximum speed of the *Shinkansen* is 200 km/h, and since it will run at high density, with a minimum of five minutes between trains, we should install a continuously operating automatic train control (ATC) system and adopt a security system in which the signal display and the brakes work together automatically. We should install the signal onboard.

> When running at high speeds, there is always the frightening prospect of an accident, so we cannot rely solely on the skills of the driver, no matter how good these skills might be. Therefore, we can imagine a system like that used in several foreign countries, where if the driver makes a mistake in applying the brakes, the signal system automatically activates them.

> ATC devices, which are already in use on domestic and foreign railways, activate an automatically controlled brake, although only when the driver has made a mistake in operating the brakes. On the *Shinkansen*, however, we would like to adopt a system in which the automatically controlled brakes are activated immediately when the train enters a speed restriction zone so that brake action is almost fully automated. This will both lighten the psychological burden on the driver while he is operating the train at high speeds and increase the precision of the brakes.

I think this explanation has given you a general concept of how an ATC system functions. If a speeding train is approaching another train from behind or approaching a station and needs to slow down, one does not have to rely on the actions of the driver because the ATC automatically applies the brakes. You might call this system one of the milestones in the long years during which signal technicians have been pursuing increased safety.

This type of system found practical application at exactly the time the *Shinkansen* was being created.

The Railway Technical Research Institute made a major contribution to the development of this ATC system. The speaker who presented the ATC concept at the lecture meeting on the plans for the *Shinkansen* at Yamaha Hall was Dr. Hajime Kawanabe, Head of the Signal Research Laboratory, who had come from the army's research institute. Dr. Kawanabe explained the disadvantages of ground signals and the principles and functions of automatic train control systems using onboard signals. He also expressed the view that ATC was essential on a high-speed railway like the *Shinkansen*. He pointed out that the *Shinkansen*, which runs on alternating current(AC), would have an interference current flowing through it that would exert a harmful influence on the signal current that flows through the rails.

For several years since 1951, an automatic operation device called 'PA135' was tested on the rubber-tired trains of the Paris Metro, but the first time in the world that ATC was actually put into service on all the trains on a line was on the Tokaido *Shinkansen*, and on the Teito Rapid Transit Authority's Hibiya subway line in Tokyo, which opened about a month earlier.

In the 30 years since the *Shinkansen* went into service, there hasn't been a single collision, and this is all thanks to ATC. During that time, the trains have run a total of 2.1 billion km, a distance corresponding to 2,700 round trips between the earth and the moon.[2]

For reasons of safety, the brakes on the *Shinkansen* are all operated automatically, but the driver operates the controls manually for accelerating after starting from a station and for stopping in the final position at another station after slowing down. Naturally, there were some arguments within the committee about whether all these operations should be automated as well.

Committee Chairman Hideo Shima responded to this by saying, "We ought to move to fully automatic operation in the future as technology advances, but we're operating under time constraints, and since we ought to stick with the options that can reasonably be implemented without excessive effort before opening the line, I would like to adopt a method that won't require us to go back to square one when we shift to automatic operation in the future."

---

[2] As of 2022, the Shinkansen has not had a single train collision in the 60 years since it began operation. During that time, the trains have traveled 5.3 billion km, which is equivalent to a round trip between the moon and the earth approximately 6,800 times.

5.2   A Maximum Speed of 250 km/h?

As I was reading this report, I noticed something odd. There was absolutely no record of any discussion of the maximum speed for the *Shinkansen*. It may have been determined in some other venue, but I wouldn't have expected the committee to decide the norms for the tracks without determining the maximum speed. At the time planning for the *Shinkansen* began, its maximum speed was supposed to be 250 km/h. At both the Trunk Line Survey Meeting and the lecture meeting of the Railway Technical Research Institute at Yamaha Hall, the *Shinkansen* trains were described as running at 250 km/h and traveling between Tokyo and Osaka in three hours (Fig. 5.1).

Yet in the records concerning ATC presented at the 19th Construction Standards Survey Committee, the figure has at some point changed to 'high-speed operation at 200 km/h.' I have been unable to find any account of the change in the official record.

During an interview, Ichiro Kato, who was in charge of the Rolling Stock Planning Department of JNR and served as the first president of the *Shinkansen* Directorate General, related the following as part of the 'unknown story of putting the *Shinkansen* into service.'

> There's one thing about the *Shinkansen* that I rather regret now. At the beginning, we made plans with the intention of having it run at 250 km/h on the current tracks, but that all went away during our negotiations with the World Bank to borrow funds. They said that they

Fig. 5.1 ATC cab signal device as seen from the driver's seat of the 0 series. *Photo* provided by the Railway Museum

wouldn't lend the money unless we reduced the speed to 200 km/h, so we reduced it, but I still sometimes find myself wondering what would have happened if we had gone ahead with our existing plans for 250 km/h.

It's really a shame that we reduced the speed to 200 km/h at the time. At least the tracks should be the original plan. We argued quite a bit about whether we should make the rolling stock and everything capable of running at 250 km/h or whether we should give up the idea. When the discussion turned to whether we should drop the speed to 210 km/h, I felt that we had thrown ourselves off a cliff, but pretty soon 210 km/h became the accepted figure, and 250 km/h became a nearly unattainable goal. When I think about it now, it really seems like a shame.

It is well-known that JNR borrowed funds for the construction of the *Shinkansen* from the World Bank. The total amount borrowed was 80 million dollars, no more than 15% of total construction costs, but it is significant that the memorandum issued at the time the money was borrowed plainly states that the Japanese government has an obligation to support the execution and completion of the construction. There was thus a clear guarantee attached that the much opposed and criticized *Shinkansen* would be built.

However, a project cannot receive financing from World Bank loans unless it is clearly proven to be technologically feasible. European and American limited express trains had been running at 160 km/h since before World War II, but that seemed to be one of the insurmountable barriers, and the common knowledge among railway technology experts was that operating a commercial service at 200 km/h would be difficult. That's why this point became a major problem when receiving financing from the World Bank.

The goals of World Bank financing are economic revival and development, and attempts at implementing unprecedented new technologies are not funded. The World Bank insisted that it could not finance any project unless it reflected 'sound engineering,' in other words, 'proven technology' that had a sufficient track record.

The Japanese explained the technical details, including various test data, and insisted that the *Shinkansen* was by no means an attempt to realize a new, unproven technology but a comprehensive combination of technologies that had already been sufficiently proved through actual use. They had evidently been able to make the World Bank understand this, so it seemed very inappropriate to press for 250 km/h.

A World Bank survey team visited Japan in May 1960, and since the 19th Construction Standards Survey Committee met the year after that, the plans must have been modified during that period.

In any case, ATC, with its maximum approved speed of 200 km/h, was unsuitable for actually running trains at 200 km/h. As the train accelerated, the brakes would be applied the moment the speed reached 200 km/h. In order to provide a bit of leeway, the maximum approved speed was raised to 210 km/h.

The entire *Shinkansen* line was equipped with Centralized Traffic Control (CTC) system. This is a device that can control the routes for trains entering each station from a central control center in Tokyo. The center is equipped with a giant train position display board, which allows staff to see at a glance where the trains are running. Railways formerly controlled the routes of trains within a station from each station's individual signal cabin.

CTC is by no means a new technology. It was introduced in the United States on a large scale before World War II, and even in Japan, the Nagoya Railroad used it on its branch lines in 1929. However, it was the first line to adopt CTC for such a long-distance lines in Japan. Each train is also outfitted with a wireless communication system that allows the control center to speak directly to drivers.

These facilities were, of course, the kinds of information systems that a railway like the *Shinkansen* needed as it was to be a modern railway, and it was absolutely essential to have accurate information of each train's position and to be able to speak directly to the driver, especially on the *Shinkansen* where the stations are so far apart. Yet as the construction of the *Shinkansen* was moving into its final stages, it became clear that construction costs were going to run over budget by a significant amount. At that point, some people, aiming to keep costs down, argued seriously that the trains would run just fine without CTC and the wireless communication system.

5.3 High-Speed Test Runs Postponed—the Plans for the *Shinkansen* that Shattered Conventional Wisdom

You could almost say that the *Shinkansen* was a reckless project from a technological point of view, because whatever else was true, no Japanese train had ever run at 200 km/h. Usually, when creating a new model of an airplane or automobile, you run endless numbers of tests before building it. I'm told that when carrying out a full-model changeover, automobile companies build more than 200 prototypes and put them through various tests until they break down.

In order to begin actual commercial train service at 200 km/h, technicians would have thought it was common sense first to confirm that their trains could hit 250 or 300 km/h in test runs, to build prototypes and test their durability by subjecting them to test runs of 100,000 or a one million km, to find out where the problems were and alleviate them, and only after that to put them into service.

In that sense, Japan's *Shinkansen* shattered conventional wisdom. Before construction work began on the *Shinkansen*, Japan's record for train speed was 145 km/h, achieved by a limited express from the Odakyu Electric Railway running on the Tokaido Line. France and Germany, too, first ran successful speed tests and only began actual operations decades later. Because the plans were laid out ahead of time and the technical tests were postponed, the *Shinkansen* was an earth-shattering project.

Behind this turn of events was the fact that France and Germany had already run tests at speeds exceeding 200 km/h and that the data from these tests were available. JNR probably also had the confidence that the *Shinkansen* had been designed using technologies that had already been put to practical use and that allowed sufficient leeway for mistakes built in.

Whatever the case may be, there is no doubt that the *Shinkansen* exceeded the expectations of conventional wisdom.

Actually, at that time, JNR couldn't have conducted 200 km/h tests if it had wanted to, because there were no suitable tracks. They had to wait until construction had proceeded for a while, and a section of the tracks had been completed.

On June 26, 1962, with tracks of a section near Odawara completed, test runs began with prototype EMU train. There was a whole list of items to be tested: whether the EMU would run safely and smoothly, whether there was any danger of its derailing, whether the ATC would function properly, whether abnormal vibration would occur between the overhead contact wire and the pantograph, giving off sparks, and how much force would be applied to the tracks.

As the experimental EMU train went through their test section, researchers analyzed the data and gradually raised the speed. Four months after beginning the tests, they achieved a speed of 200 km/h, and on March 30 of the following year, they recorded a speed of 256 km/h. In order to run a 200 km/h commercial service on a daily basis, they would have to achieve at least this speed during test runs in order to confirm that they would have some room to maneuver. Evidently, the train was able to maintain that high speed for only about 10 seconds, but after the tests were over and the researchers examined the section of track where they had taken place, they found that the rails were warped. The bogie of the EMU had begun to hunt, and that force had warped the rails.

Even though everyone was glad that the test trains had not derailed, they could not run at 250 km/h on a daily basis under these conditions. It may have been a good thing after all that JNR lowered the maximum speed based on the World Bank's opinion.

This 32-km test section was referred to as the 'model line,' and it was used over a period of two years to conduct a total of 250,000 km of test runs with three-car EMU trains. During that time, 150,000 visitors took test rides, including the Imperial family and the astronaut, Colonel John Glenn, who was more recently in the news for traveling on the space shuttle at the age of 77.

There seem to have been several unusual episodes during the tests runs on the model line. Shigeru Otsuka, who was the engineer in charge of the tests at the time, told me of the following incident.

On the day before President Sogo was supposed to attend the formal beginning of the test runs, the researchers tried running a test EMU train just to see what would happen. As the train approached the end point, its weight caused the ground under the tracks to sink, and the pantograph lost contact with the overhead contact wire. They hurriedly lowered the wire and somehow managed to extricate the train from its difficulties. The problem was that the earthworks under the hastily constructed tracks had not had time to harden.

In addition, once the tests began, birds such as pigeons and sparrows frequently flew into the front of the EMU train. However, as the year progressed, these incidents became less and less frequent, as if the birds had become aware of the danger. The same type of thing happened after the *Shinkansen* went into actual service. It may be that the birds' instinct for predicting danger was transmitted to other birds as well.

On July 25, 1964, all the rails were finally connected, and over a period of 10 hours, the first slow non-stop test run of the entire line was conducted. The test ended without incident, but near Lake Biwa, the train ran onto a gas cylinder that someone had accidentally left on the track during construction and sent it flying. Trailing gas behind it, the metal cylinder soared into the air like a rocket.

One month later, on August 25, JNR conducted a public test run that would cover the distance from Tokyo to Osaka in four hours, the time it was supposed to take on its actual service schedule. NHK followed what was happening on every section of the entire line and presented live, on-the-spot broadcasts. That's how much of a public event the opening of the *Shinkansen* was.

Reisuke Ishida, the new president of JNR, rode along on this test run, and since I was in charge of operating facility planning at that time, I, too, was able to ride along. Surrounded by ranks of television cameramen, I was in a feeling of beaming joy, but what impressed me most was the number of people who came out to see the first run of the *Shinkansen* for themselves. There were people, people, and more people everywhere—on top of office buildings and apartment houses, along riverbanks, and on top of the road overpasses that crossed the parallel Tokaido Line. I can still remember the amazed expressions on the faces of some young women looking up at the raised track near Nagoya as they did their laundry.

Several helicopters were flying along with us as part of the newsgathering works. One of them descended to the same height as the test train and tried to fly alongside it at a low altitude to take pictures, but the *Shinkansen* was faster and passed the helicopter. This was a fresh and surprising revelation to me.

The test train arrived at Shin-Osaka safely, and on the return trip, I visited the driver's cab.

In the assistant's seat, next to the driver, who was nervously gripping the control handle, sat Shigeru Otsuka, who had been in charge of the model line administrative depot. He was telling the driver what speed to run the train at, and that speed changed every few kilometers. Since the earthworks under the newly built tracks had not yet hardened, there were many places where it was impossible to go very fast. In particular, as we approached bridges, it felt as if the train was creeping onto the structure. Since the foundations of the bridges had been reinforced with concrete, they didn't sink, but the earthworks under the tracks before and after the bridges did.

Under these circumstances, it was impossible to run at 200 km/h. Otsuka had a chart indicating the speed limit established for each section, based on a detailed inspection of the tracks, and he was using it to indicate the proper speed to the driver. As it was a non-scheduled test run, it wasn't operating on automatic train control. It was operating on Otsuka's own Human Train Control.

However, only one month remained between this test run and the formal opening of the *Shinkansen*. Some people on the test train were of the opinion that it would be impossible to begin operations at 200 km/h and that it would be better to reduce the speed. Here is what Ichiro Kato, the first President of the *Shinkansen* Directorate General, said about what went on at that time.

There were strong opinions in favor of reducing the speed to 160 km/h, not that anyone thought that 200 km/h was dangerous, of course, but... Well, at that time, I wanted the maximum speed to be 200 km/h, even if the train reached that speed for only a short time. At other places, it would be fine if it ran at 160 km/h. It would be simple to arrange that with the ATC, and I just wanted them to give up the idea of reducing the maximum speed. I asked, and they accepted my request.

At the time, I felt that if we made 160 km/h the maximum speed, we'd have to go through another stage of raising it to 200 km/h, and I determined that this would probably be a lot of bother.

We would start up in 1964 with the *Hikari* making the trip in four hours and the *Kodama* making the trip in five hours, and within a year, when we lowered the times to three hours and 10 minutes, and four hours, people would get the impression not that we had compressed our running hours but that we had speeded them up. I explained that this is why we wouldn't increase our speed. We would simply decrease the number of places where the speed was restricted. With that, they finally got what I was trying to say. I think that if we had raised the speed limit by 40 km/h, we never would have been able to lower our traveling hours to three hours and four hours, so now I'm glad that we persevered.

Thus the *Shinkansen* formally went into service on October 1, 1964, with the *Hikari* traveling between Tokyo and Shin-Osaka in four hours, and the *Kodama* making the trip in five hours, and on November 1 of the following year, the running time became three hours and 10 minutes. It seems almost miraculous that this massive construction project, which was also the world's first high-speed railway, took only five years to complete, including the necessary technological developments (Fig. 5.2).

Fig. 5.2 JNR President Reisuke Ishida at the ribbon-cutting ceremony for the *Hikari* 1 as the Tokaido *Shinkansen* begins operations. October 1, 1964, at Tokyo Station. *Photo* provided by the Railway Museum

The Tohoku *Shinkansen* and the Joetsu *Shinkansen* took 11 years to complete. They also went into operation at the provisional starting point of Omiya, because construction of the line into central Tokyo wasn't completed in time. Undoubtedly, the Japanese people had high hopes for the Tokaido *Shinkansen*, and it came about due to the enthusiasm and efforts of the people involved, but that's not all. According to former JNR president, Iwao Nisugi, who was directly involved in the construction of the *Shinkansen* as Head of the Nagoya Trunk Line Construction Bureau, "Deregulation had a great effect."

He wasn't speaking of bureaucratic regulations but of the fact that a nearly independent organization, the *Shinkansen* General Bureau, was created and allowed to determine its technological standards and construction plans freely and that a broad range of rights was delegated to the various Trunk Line Construction Bureaus.

We were able to get it done because we didn't have to ask the Headquaters about every little thing, Nisugi told me.

JNR Headquarters was second to none in issuing regulations. This may be something like the problems we have with the current administrative reforms, devolution of rights to local governments, or the privatization and breakup of JNR itself. As might be expected, there were strident criticisms and expressions of dissatisfaction from the existing departments and bureaus.

> The *Shinkansen* General Bureau is like the Kanto Army in the old Japanese military, someone said with exaggerated displeasure. "They're loose cannons, doing whatever they want without consulting anyone." Some officials grumbled, "Old Man Sogo is letting the *Shinkansen* General Bureau run wild.

As a staff member of an existing department myself, I could understand very well what they were saying.

By the time the *Shinkansen* was a reality, it was the result of the zeal and efforts of a large number of people, not just the construction departments. The engineer Shigeru Otsuka, who served as an advisor for the full-length test runs of the *Shinkansen,* was one of these. Akira Hoshi, one of Japan's foremost experts on rolling stock technology, contributed to reducing the weight of Japan's rolling stock and to the development of the *Kodama* limited express, and he told me recently, "The person who contributed the most to the development of Japan's EMU trains was Shigeru Otsuka."

Soon after joining JNR, Otsuka worked as an inspector in the Mitaka EMU Depot. In his role as technical expert, he advised older inspectors on how to deal with JNR's first new performance train, the 101 (then the Moha 90). After that, he served as assistant manager at the Tamachi EMU Depot, where he was responsible for the inspection and repair of the new *Kodama* limited express.

Later, he moved on to the Moji Railway Administration Bureau in Kyushu, where he was in charge of preventing breakdowns of Japan's first AC-DC train because the high-voltage train was constantly buffeted by salty winds on the northern coast of Kyushu. This damaged its insulation and led to continual breakdowns. When tests of the *Shinkansen* began on the model track at Odawara, Otsuka moved there also, this time to lead the high-speed tests.

When a big new project like the *Shinkansen* succeeds, the technical experts who design and manufacture the rolling stock and facilities are not the only ones who deserve credit. The experts who master the new technology, overcome the breakdowns that are an inevitable part of any new system, and acquire the know-how to improve and maintain the technology also play an essential role. Japan's ability to foster these kinds of human resources was one of the great strengths that led to the development of the *Shinkansen.*

5.4 An Unprecedented Type of Pattern Scheduling

Once it opened, the *Shinkansen* proved to be a unique railway, the only one of its kind in the world, running nothing but limited express trains for 30 round trips each day. It also only had two types of trains, the *Hikari* and the *Kodama*. Every day, at the top of the hour, a *Hikari* left Tokyo Station, and at half past the hour, a *Kodama* headed out. Since this pattern of departures was repeated each and every hour throughout the day, JNR called it 'pattern scheduling.'

Some urban commuter EMU train lines also ran on this kind of schedule, but this was the first time in Japan, perhaps even in the world, that a long-distance limited express EMU train running throughout the day had adopted this kind of schedule. In doing so, it provided a model for a new kind of scheduling.

Until the *Shinkansen* came onto the scene, limited expresses were literally 'limited' trains. They ran only a limited number of times per day and only when there were likely to be a lot of passengers. Even on the Tokaido main line, Japan's number-one trunk line, the limited express ran more than 10 round trips per day except for the overnight Blue Trains, which had an exceptionally high number of limited express trains.

Strictly speaking, the Tokaido *Shinkansen* was not the first line to run express trains at fixed, equal intervals. I recall that soon after I began working for JNR, I read a newspaper article about Dutch Railways (NS) running express trains at fixed, equal intervals to its major cities. Recently, I sent an inquiry to NS and received the following reply:

> The answer to your question about 'equal interval' trains is that this kind of scheduling began on May 15, 1934. We ourselves were quite surprised to find out that this had started in the 1930s. As you can see from the enclosed copy of a schedule from that time, there was service between Amsterdam and The Hague, between Amsterdam and Utrecht, between The Hague and Rotterdam, and also across the border into Belgium every thirty minutes.

The distance between Amsterdam and The Hague or Utrecht is measured in tens of kilometers at the most, and there was no surcharge for this service, so, like the *Acty* on the Tokaido Line or the New Rapid Service in the Kansai area, it was more of a rapid transit service than a limited express. In that sense, we may say that the

Tokaido *Shinkansen* was the first long-distance limited express to run on a pattern schedule.

The situation was the same on European railways when I first visited Paris in 1966. France's closest counterpart to the *Shinkansen*, the Paris–Lyon Express, had several trains leaving Paris or Lyon around nine o'clock a.m., then nothing until a single train at noon, and then no limited expresses again until evening. Pattern scheduling of express trains between major cities gradually became more common in Europe in later years, perhaps spurred on by the success of the *Shinkansen*.

The first national railway to try it was British Railways (BR). Taking the opportunity afforded by the complete electrification of the main west coast line from London to Liverpool and Manchester, 300 km away, British Railways' first 160 km/h, pattern-scheduled limited express trains made their appearance on April 18, 1966, a year-and-a-half after the completion of the Tokaido *Shinkansen*.

As part of the publicity campaign for the new service, British Railways developed a series of television programs about the "Inter-City Heart-to-Heart," profiling 40 sightseeing spots along the route (Fig. 5.3).

The campaign resulted in a 66% increase in ridership. Since then, 'InterCity' has taken hold as a sort of brand name for a service that links major cities at equal intervals. The name is not limited to Britain but has spread to Germany, Switzerland, and Denmark.

In 1971, West Germany began a new 'InterCity' (IC) service that ran limited express trains made up entirely of first-class coaches between its major cities every two hours. Germany has no single dominant urban center comparable to Tokyo or London and instead has a number of mid-sized cities scattered throughout the country.

Fig. 5.3 Long-distance train operated by Virgin Trains East Coast with a maximum speed of 201 km/h. In 1994, British Railways was privatized, and the name 'InterCity' is no longer used officially. *Photo* provided by Tadashi Sumita

The IC limited express travels on several fixed routes, and stations, where these lines intersect, have platforms where trains from both lines stop at the same time on either side. This allows passengers to transfer freely from one line to another.

Swiss Federal Railways (SBB/CFF/FFS) instituted a similar type of service, also called the 'IC,' and also running at fixed intervals on a so-called Takt-Fahrplan ('rhythm-of-a-pendulum schedule').

Denmark also began IC service in 1974, but since almost none of its lines were electrified, its flagship trains all ran on diesel fuel, and the new DMU cars that a well-known railway designer named Jens Nielsen designed for this line generated a lot of comment. Denmark's many islands were linked by a system in which trains rolled directly onto ferries and traveled at fixed intervals between the capital, Copenhagen, and such cities as Odense and Aarhus. In 1997, a combination bridge and tunnel over the Great Baelt Straits in the Baltic Sea was completed at about the same time as the electrification was completed, so now a new EMU IC service is in operation. I think it would be interesting to visit Hans Christian Andersen's hometown of Odense, traveling through the tunnel and on the bridge.

Like the *Shinkansen*, the InterCity expresses marked the beginning of a revolution in the limited express train. Rail systems had once been made up of special trains that ran only at certain limited times. Now they were made up of conveniently scheduled trains that a person could ride at any time. But the downside to all this was that the individual trains lost their distinctive character. Express trains became more accessible to the masses at the expense of their individuality.

5.5 The End of the Era of Distinctive Express Trains

Until these developments took place, the typical express train was like Japan's *Tsubame* or *Hato,* Britain's *Flying Scotsman,* or France's *Le Mistral,* a train with its own name and features found on no other train. Even now, some of Germany's ICE limited expresses and EC international limited expresses have individual names, but one would be hard-pressed to say that they have individual features or public images. Changing times have brought about more of a desire for convenient service of even quality than for nostalgia or individual charm.

As many sources record, Japan's first ordinary express train began operating on the Sanyo main line between Kobe and Hiroshima in 1894, when it was still a private railway. The 'fastest express,' the previous incarnation of the limited express, began running between Shinbashi and Kobe in 1906, and it was also the first express train to introduce the practice of levying an express surcharge on passengers, a practice that continues on Japanese railways to this day.

The appellation 'limited express train' was formally introduced in 1912, and it referred to a train that ran from Shinbashi to Shimonoseki in 25 hours, eight minutes, a trip that now takes about five hours, 30 minutes on the *Shinkansen.*

The distinction between ordinary expresses and limited expresses is unique to Japan, and the way trains are classified varies from country to country. The most common term is 'express,' which corresponds to Japan's ordinary expresses but may also include limited expresses. The famous *Orient Express* is simply an express train. There's nothing about its being 'limited' in its name. Even in Japan, you see fewer and fewer ordinary express trains, so in fact, the European express and the Japanese limited express can be considered equivalents.

Evidently, railways have always had trains that stopped at fewer stations and ran faster than ordinary trains. As you probably know, the world's first real railway began running between Liverpool and Manchester, England, on September 15, 1830. Its first train took four-and-a-half hours to travel the 50 km between the cities at an average speed of 11 km/h.

This railway did not mix first-class and second-class coaches on the same train but ran separate one-class trains, supposedly because the English aristocracy disliked the idea of riding the same train as uncouth second-class passengers. The first-class trains were made up of coaches that looked like horse-drawn coaches on rails and were painted yellow, hence the name 'Yellow Train.'

The second-class coaches looked like open-air freight cars with wooden benches and were painted blue, which gave rise to the name 'Blue Train.' The second-class train stopped at every station along the route, while the Yellow Train stopped at only one station between Liverpool and Manchester, which means that it was a kind of 'express.' Of course, it wasn't called an express, but the idea of a faster option for train travel was born at the same time as the idea of passenger rail service itself.

It's not clear exactly when the word 'express' was first used, but it seems to have gained currency in England and France in the 1840s.

The world's first express train with its own name was the *Irish Mail,* which began operating between London and Holyhead, Wales, in 1848. It got its name because it was part of the process of transporting mail from England to Ireland.

At first, the *Irish Mail* was merely a nickname bestowed by passengers out of fondness for this fast train, but the railway company later recognized it officially and made it the formal designation. The trains that transported mail in Europe were the fastest ones in service at the time, and the phrase *mail train* brought to mind a high-quality train.

The Belgian company Wagons-Lits created the first real limited express trains in 1878. This company manufactured sleeping cars, dining cars, and salon cars that were extremely luxurious for their time. They attached these cars to trains operated by different railways and let these railways use them for a special fee. Wagons-Lits provided the cars and the crew members and supervised the services offered in the cars. At that time, Europe's rail network was operated by a multitude of small, private companies, and Wagons-Lits had a monopoly on the business of providing cars for long-distance travel among several railway companies.

At that time, long-distance travel was limited to the wealthy, and Wagons-Lits began offering high-priced, deluxe tours to such people. Wagon-Lits put together trains for this purpose called *Grands Express Europeens* (Great European Expresses), which were made up entirely of luxury-class cars.

The first of these *Grands Express* was the *Orient Express*, which began running between Paris and Istanbul in 1883 (Fig. 5.4).

Fig. 5.4 Orient Express train by Wagons-Lits at Shinagawa Station. *Photo* provided by the Railway Museum

However, Germany was not pleased with the monopoly of Wagons-Lits, which was based in France and Belgium. In 1916, right in the middle of World War I, a German company called Mitropa began offering services similar to those offered by Wagons-Lits, particularly in Germany. Competition between the two companies is said to have been quite fierce.

However, as airplanes developed, fewer and fewer people chose this luxurious but time-consuming form of travel, and even Wagons-Lits began losing money on its train-related business. In the 1970s, it decided to get out of the railway business, and at present, all it does is provide dining car service on a few trains.

Unlike French trains, which almost never have dining cars anymore, most German express trains still have dining cars, and Mitropa continues to offer meal service, but the food is not up to the level that would make the meals enjoyable experiences in themselves.

As the InterCity trains went into service, the age of the luxurious limited expresses faded into history.

5.6   Reserving Seats: The Curtain Rises on the Computer Age

These days, you can easily reserve seats on any JR train in Japan at any time at the so-called Green Reservation Offices[3] in your local train station. It wasn't always like this.

In order to ride any train that required reservations, such as the *Tsubame* limited express, you had to have a station staff phone the reservation center in Tokyo. The line was always busy, which is not surprising when you consider that the reservation center had only a little over ten staff taking reservations from all over Japan.

The reservation center inside the red brick walls of Tokyo Station contained a big round table, in the center of which was a rotating tray of the type seen in Chinese restaurants. On top of it lay the reservation rosters for each train. When a request came in by phone, the ticket agent took the reservation roster for the relevant train off the rotating tray and added the reservation if the train had seats available. The agent then tossed the roster back onto the rotating tray, which revolved constantly at a fairly high speed. The scene inside the reservation office resembled a gathering of sleight-of-hand artists.

Since only limited expresses, sleeping cars and special second-class seats (the present-day 'green car' seats) on some of the ordinary expresses required reservations in those days, the agents were able to keep up more or less, but when the number of limited expresses such as the *Kodama* increased, the system became unworkable. In 1960, therefore, JNR developed and began using the so-called Magnetic Electronic Automatic Reservation System (MARS) computer system.

Data entered at such major stations as Tokyo, Shinjuku, or Osaka was sent directly to a central computer in Tokyo Station that could immediately issue reserved seat tickets. It was the very beginning of the computer age, and MARS was Japan's first real-time online system.

It was a small system, however, quite unable to handle the reservations for all reserved seats. Instead, the MARS-1, capable of processing only 3,600 reservations per day, simply handled the reservations for the four *Tsubame* and *Kodama* trains.

Those were precisely the years in which the number of limited express runs was increasing, and with them, the number of potential reserved seats. The small computer system simply couldn't handle the job, and JNR set about developing its next fully functional, large-scale computer system.

When the *Shinkansen* first went into service, reservations for it were not computerized. JNR officials thought that because of the large number of trips per day, passengers could just reserve their *Shinkansen* tickets at the station where they planned to board. People in those days didn't know much about computers, and they probably assumed that the old system with the reservation rosters would work just fine. It's often difficult for people to change the procedures that they've grown accustomed to over the years.

The result was a fiasco. It took hours to reserve a seat, and many people were unable to get on the train they were aiming for. Newspapers the day after the line opened

---

[3] Midori no Madoguchi in Japanese.

carried such headlines as "The *Shinkansen*: Advanced Sales in Chaos." Here's how one article reported on the confusion: "The Tokaido *Shinkansen* got off to a much-hailed smooth start, but sales of passenger tickets were thrown into chaos due to the ticket agents' unfamiliarity with the procedures and breakdowns of the newly installed machines. All day and into the night, some passengers at the advanced sales counter at the Yaesu entrance to Tokyo Station shouted that it was taking five hours to buy tickets. Unable to handle the mad rush of people, the agents stopped taking applications temporarily, and the whole scene was a mass of confusion. Finally, the Railway Police Staff went into action and restored order."

JNR hurried to create a reservation system for the *Shinkansen*, and the following year, as the *Hikari* began running between Tokyo and Osaka in three hours and 10 minutes, the *Shinkansen* reservation system was computerized.

The new MARS system's central mainframe computer was linked to terminals in the Green Reservation Offices in 1,550 JNR stations nationwide and 7,300 travel agencies. It sold not only reserved seat tickets but also hotel vouchers and tickets for concerts and events.

Each day, 2.7 million requests come into the computer through the terminals, and it issues 1.6 million reserved seat tickets. Now that airline reservations are online, you can buy airplane tickets at the Green Reservation Offices, too.

Fig. 5.5 'Green Reservation Office' at Shinjuku Station, where people can reserve seats on all JR trains. Many automatic ticket vending machines are now in use. *Photo* provided by Tetsuro Aikawa

MARS was Japan's first real-time online system, and in its present form, it is comparable in size to nationwide bank automatic teller machine (ATM) systems, which have about 4,500 terminals, or to the inventory management system for the Seven-Eleven convenience stores with about 7,200 terminals, which puts it on nearly the same scale as the MARS system. These systems all have different functions, but they are similar in scale (Fig. 5.5).

Although the MARS system was the world's first computerized seat reservation system for trains, airlines had set up such systems earlier. American Airlines contracted with the Teleresistor Company to build a computer, especially for making plane reservations, in 1953, when computers had just begun to find the practical application and still ran on vacuum tubes and relays. By today's standards, this computer was a primitive machine, but it was able to handle 12 days' worth of reservations for 1,000 flights on 100 terminals.

In 1963, a full-scale system for reserving airline seats, the famous SABRE system, appeared, and it was a general management information system (MIS) capable of not only reserving airline tickets but also processing all sorts of information services, data on individual customers, and every other type of data that is useful from a business perspective. At present, it consists of 130,000 terminals in 70 countries, serving 30,000 travel agencies and 35,000 hotels throughout the world. It sends 200 million messages per day.[4]

It's one of the world's largest real-time online systems, second only to the U.S. government's system, and one can also connect to it from a personal computer via the internet. SABRE processes travel services worth about 40 billion dollars per year, about 2.4 times the amount processed by the MARS system.

5.7 Continual Accidents and Problems

The Tokaido *Shinkansen* was a huge hit. Shortly after it opened, it carried 60,000 passengers per day, but two years later, that figure had doubled. In 1975, 10 years after the running time between Tokyo and Osaka had been decreased to three hours and 10 minutes, the passenger volume had reached 320,000 passengers per day, nearly six times the number it carried at the outset.

In that year, the total distance traveled by passengers using the *Shinkansen* was 35.2 billion passenger kilometers, 1.4 times higher than that projected during the planning stages of construction.

The number of trains also quadrupled during those 11 years, so there were now 130 roundtrips per day. Originally, each train was made up of 12 cars, but the *Hikari* was increased to 16 cars at the time of the 1970 Osaka World Exposition.

---

[4] Sabre now serves customers in more than 160 countries, including 350,000 travel agencies, 450 full-service and low-cost carriers, and 1.3 million hotels worldwide. Revenue totaled USD$1.7 billion for the entire year of 2021.

These days the Tokaido *Shinkansen* makes 140 round trips per day and carries 370,000 passengers. It's very unusual for a project to succeed this well.[5]

However, I would never claim that the *Shinkansen* had a trouble-free beginning from a technological point of view. All one needs to do is look at newspaper articles from that time.

On October 2, 1964, the day after service commenced, one newspaper carried a headline that read "Trouble Again on the *Shinkansen*." It told how part of the overhead contact wire near Toyohashi had slackened so that 10 or more trains were delayed for 15 minutes.

Next, the headlines on November 23 read, "First Major Accident on the *Shinkansen*." The article told how five members of a track maintenance crew were hit by the first *Kodama* of the day and killed instantly.

Problems continued after that. On December 16, the first accident related to the overhead contact wires stopped trains near Shin-Osaka for two hours. Two days later, more trouble occurred, described in a headline that read, "Is This the Cream of JNR's Technology? A Stopping Train Journey on a Super Limited Express. Passengers Given the Runaround in the Midnight. This was another mishap with the overhead contact wires, and it shut down the system for five hours.

The following year, we find headlines reading "The *Shinkansen* Can't Take the Snow" and "Tokyo Station a Swirl of Anger and Confusion after a Double Accident" (both January 6), "*Shinkansen* Stuck Again" (February 6), "Power outage Disrupts *Shinkansen*" (March 26), "This Morning's Earthquake Throws the *Shinkansen* into Confusion, Causes the Roadbed to Sink in eight Places" (April 20), "*Shinkansen* Sees Major Disruptions in the Evening During the Spring Holiday Period" (May 2), "The *Shinkansen* Roadbed Collapses Again. Schedules Disrupted in two Places Due to Rain, Passengers Demand to See JNR President" (May 3). It continues with "The *Shinkansen* with All Its Weak Points: the Roadbed Sinks in 17 Places." These are all events that happened within little more than half a year after opening, and all sorts of minor problems occurred in addition to these.

I myself had the experience of being confined to a *Shinkansen* train for several hours as a passenger at the time of the May 1 disruption due to the overhead contact wire trouble. That day, I went to Nagoya on some private business, and I had originally intended to return on the nighttime semi-express, *Tokai*, which ran on the conventional Tokaido line. However, since we were scheduled to move to a new house the next day, I decided that I'd prefer to get home before the day was over, so I paid the high express surcharge and transferred to the *Shinkansen*. That turned out to be a big mistake.

I'm pretty sure that I transferred to the *Hikari 26*, which left Nagoya at 8:28 p.m. Things seemed to be going smoothly as we sped out of Nagoya, but around Toyohashi, the train kept stopping and starting repeatedly until it stopped dead just before Hamamatsu Station. Unfortunately, we were stalled on a curve, so the train was leaning to one side. You don't feel this leaning motion when the train is moving

---

[5] As of 2019, the Tokaido Shinkansen line had 189 round trips per day, with approximately 460,000 passengers.

at high speed, but once it stops, you can see that it leans at quite an angle. When I tried to open the passage door to go to the toilet, I had to apply a considerable amount of force before it would budge, and it slammed shut behind me the moment I let go of my hand.

Since this was a mishap involving the overhead contact wire, all the lights except the emergency lights went off. The ventilation equipment also stopped working, and the temperature inside the train kept rising. We somehow managed to pinch something to keep the heavy passage door open and manually opened the entrance door in order to let outside air enter.

Since we were very close to Hamamatsu Station, I thought of jumping off the train and walking there, but *Shinkansen* trains are surprisingly high off the ground, and I didn't think I could jump out without injuring myself. During that time, the night train *Tokai,* the train that I should have been on, passed by us on the track below. We sat there for about five hours before the train started to move. By the time we came into Tokyo Station, the sky was already growing light.

As I said before, this was a mishap involving the overhead contact wire, which extends above the track, is supported by utility poles, and supplies electricity to the train. On curved sections, the curves on the track and the curves in the overhead contact wire system have to match, so the wires are strung between the poles and held in place by metal fixtures called 'steady rest arms.' What had happened in this case was that the arms had come loose and were hanging down when the pantographs of the speeding train collided with them and broke.

The reason for the stoppage was that a train, a few trains ahead of the one I was on, had broken down near Shizuoka Station. As soon as this incident occurred, a JNR employee was rushed to the scene to figure out what had happened. Since the train had six pantographs, they all had to be replaced, and since the overhead contact wire had wrapped itself around the pantographs and snapped off, a new wire had to be strung. As a result, repairs took about four hours.

The *Shinkansen* had a lot of trouble with mishaps involving overhead contact wires and sinking roadbeds. In incidents reminiscent of the 1955 SNCF incident in which the pantograph of an experimental train burned off at high speeds, 'current collection' (taking in current through the pantograph) was one of the biggest sources of trouble.

Anyway, when a train is speeding along at 200 km/h with metal surfaces coming into contact, a huge 200A current is flowing, so when vibration causes the pantograph to separate from the contact wire, there are not only big sparks but also a large amount of shock to the steady rest arms. Of course, the design of the system was determined during the construction planning phase after a lot of studies and tests, but once you put a system like this into commercial service and actually many trains start running every day, it is not unusual to get results that are at odds with the plans. Sometimes troubles, that we never anticipated, occurred.

These problems were called 'initial breakdowns,' and the Tokaido *Shinkansen* was plagued with them, as you might expect from a system that was planned, constructed, and completed in a mere five years. The most common problems involved overhead contact wires, broken rails, and the sinking of the track bed due to rain. In fact, the system was harshly criticized for being unable to stand up to rain.

However, within two years after the line opened, the series of 'initial breakdowns' dwindled away to nothing, and the Tokaido *Shinkansen* was able to operate with everything in order and working correctly. Before a new technology can really be put to use, we not only have to develop the technology, design and manufacture it, and conduct endless tests—we also have to go through the essential step of working through all the initial breakdowns. This is particularly true of railways, which get by with an extremely small number of prototypes and tests.

Unlike automobile manufacturers, who may make hundreds or thousands of prototypes for a new model, JR doesn't have its own test course, so tests with a few short test trains and struggles with the initial breakdowns determine whether the technology will survive or not. It was, therefore, only natural for the *Shinkansen* to be troubled by initial breakdowns, and the reason that it was able to operate safely and on schedule in a relatively short time is the 'sound engineering,' proven technology, advocated by the World Bank.

5.8 Snow and Noise

Once JNR had overcome the initial breakdowns on the Tokaido *Shinkansen*, three further trials awaited it. They involved problems with airtightness, snow, and noise.

The first problem was one that JNR had completely failed to anticipate. Ever since the test runs at Odawara, they had known about some problems with airtightness. They found that when a speeding *Shinkansen* entered a tunnel, the abrupt change was enough to make passengers' ears pop. The design for the mass-produced EMU cars was hurriedly changed to make them airtight so that the vents in the ventilation system would close just before a train entered a tunnel automatically.

But since the design was modified so hastily, the passageways between the cars and the toilets located there were not airtight. For that reason, during the initial period of operation, there were several dreadful incidents in which passengers who happened to be using the toilet when the train entered a tunnel ended up covered with waste from head to toe.

The cars were then remodeled to be completely airtight. This airtight structure is the reason that there is a short interval before the doors open after a *Shinkansen* has stopped at a station. You hear a brief sound, and the doors seem to be moving slightly outward.

The *Shinkansen* was also one of the first trains to have only septic tank-type toilets. I hesitate to mention this unpleasant topic, but the toilets on ordinary trains in those days dumped their waste directly onto the tracks. In the old days, there was no alternative, so everyone put up with the situation, but as the standard of living rose, there were increased complaints about 'yellow pollution' from people living along the lines and from JNR employees who inspected and maintained the tracks. The toilets on the non-*Shinkansen* trains were therefore also gradually remodeled and equipped with septic tanks. Yet in Europe, there are still many trains, aside from the TGV and ICE, in which the toilets flush directly onto the tracks.[6]

The problems with snow were extremely serious. The first time they reached crisis proportions was on December 16, 1965, less than two years after the beginning of operations.

On that day, a fierce Siberian cold wave blasted the Japanese islands, leaving the Maibara area near Lake Biwa buried in 40 cm of snow, and snow plow cars were run on the *Shinkansen* Line as well as other conventional lines. The day's first Tokyo-bound train, the *Hikari 2*, passed through Maibara, but just as the train was approaching Sekigahara, the section with the most snow, it broke down and wouldn't go any farther.

That wasn't all. The next train also broke down as it moved into the area of heavy snow. Inspections revealed that the control and braking apparatus under the railcar were damaged in several places and that dozens of EMU windows were cracked. After three hours, crews somehow managed to move the damaged EMU trains to stations at Gifu-Hashima and Nagoya. In the end, more than half of the *Shinkansen* runs that day were canceled.

The cause of these problems was snow. When the *Shinkansen* ran through areas of deep snow at 200 km/h, it plowed up a spray of snow with tremendous force. Viewing videos of a train running under these circumstances, researchers saw that it was enveloped in a cloud of snow, which adhered to the various devices underfloor of the EMUs and to the bogies in the form of ice. When these lumps of ice fell off, they caused the ballast around the tracks to fly, breaking the devices on the underfloor of the EMU train and some of their windows.

This was a completely unanticipated occurrence, and it taught us about the dangers of traveling at 200 km/h. It was not unusual for limited express trains to run through areas of heavy snow on conventional lines, but none of them had ever experienced these severe problems. After these problems occurred, trains not only slowed down to between 160 and 170 km/h after heavy snowfalls, depending on circumstances but also stopped at Nagoya Station so that station crews could chip the ice off the devices under the floor of the EMUs. This was because temperatures tended to be warmer beyond Nagoya, making the ice more likely to fall off (Fig. 5.6).

Chipping the ice off by hand sounds like an exceedingly primitive method, but we were unable to find any other method for solving our problems with snow. We subsequently tried a variety of methods, including blowing the ice off with compressed air, melting it off with infrared heat, and covering the devices with a special ice-resistant coating, but none of them worked very well. The most effective method was spraying water on with a sprinkler, the same method used on highways in snowy regions. It was just that experts worried that the earthworks under the *Shinkansen* tracks would soften and collapse, unlike the pavement on a highway, if too much water were applied.

---

[6] More than 20 years have passed since this book was written, so the number of railcars with controlled emissions toilets has increased since that time.

Fig. 5.6 Chipping off ice and snow that has adhered to the *Shinkansen* EMUs. Jan. 1984 at JR Nagoya Station. *Photo* provided by Kotsu Shinbunsha

They did their best to make sure that the snow would not fly up, and even nowadays, the trains are forced to go slowly during heavy snowfalls. Crews still remove the snow at Nagoya Station, but the primitive method of chipping it off with a stick has been mostly replaced by blasting it off with high-pressure steam. On the Joetsu *Shinkansen,* the underfloor devices are body mounted and covered, and the tracks are built not on earthworks but on a totally concrete structure. When it snows, large amounts of water heated to a temperature of 10 °C are sprayed to melt the snow. This has allowed the Joetsu *Shinkansen* to avoid schedule disruptions, despite its route through a region of extremely heavy snowfalls. Of course, these kinds of preventative measures cost a lot of money. A concrete structure is more expensive than a roadbed situated on earthworks, and the cost of just heating the water to melt the snow and spraying it with the sprinklers is 1.3 billion yen per year.

Another major problem that the *Shinkansen* faced was that of noise and vibration. The *Shinkansen* was a great success, both technologically and in terms of the service it offered. Many people praised it as a symbol of Japan's high level of railway technology, but there was a downside: noise and vibration. Even before the *Shinkansen* was built, there had been noise along rail lines in large cities, but now, JNR had built a new, high-speed railway with a large number of trains in areas where there had been almost no trains before. Therefore, noise became an increasingly serious problem, especially in areas with a lot of housing.

There were complaints about noise and vibration from the time the *Shinkansen* opened, but they accelerated and intensified in the 1970s. Behind this new spate of complaints was the fact that pollution, including air and water pollution, was receiving a great deal of close attention as a social problem, perhaps in reaction to the recent rapid economic growth. In 1972, the Environment Agency urged JNR to take measures against noise, and in 1975, environmental standards for noise were established for the *Shinkansen* after much discussion.

Noise and vibration were extremely difficult problems to solve technologically. No one could, in fact, find any technology for reducing the noise level. However, there have been continued efforts to reduce the effects of noise and vibration by building anti-noise barriers and improving the rolling stock. There have also been projects to soundproof houses in particularly noisy areas.

Ever since the *Shinkansen* went into service, an amount adding up to 170 billion yen has been invested in environmental measures against such nuisances as noise and vibration. At present, the noise level is less than 75 decibels in most areas, but further efforts will be needed.

If you're wondering how much the noise level has been reduced, I can tell you that the noise level of a train traveling at a given speed is now four decibels lower than it would have been when the *Shinkansen* first opened. You may be thinking 'Only four decibels?' but a reduction of four decibels means a 60% reduction in noise energy.

5.9   Trouble on the *Shinkansen*

After the *Shinkansen* had overcome its initial problems and run fairly well for about 10 years, problems once again seemed to be popping up all over and in all parts of the train. Not only did the mechanical parts of the rolling stock break down more often, but cracks began developing in the car body, making it less airtight.

When two trains passed in opposite directions inside a tunnel, it sometimes seemed as if the car bodies were being dented. As these situations continued over ten years, metal fatigue set in, and that's when the cracks appeared. Nothing like this had ever happened on conventional lines.

This is off the subject, but the first *Shinkansen* EMUs manufactured began to be replaced with new EMUs in the 13th year of operation. They were replaced not because the parts necessary for running safely were damaged but because the car bodies were old and were losing their airtightness. Because of that experience, all the structural parts of the EMUs are reinforced, so they will have longer service lives.

There were also more and more incidents in which rails broke. It was during this period that we began replacing the rails on the entire line and putting down fresh ballast. Breakdowns in the power facilities and the signal and safety facilities also occurred more frequently. Incidents in which overhead contact wires broke, stopping all trains for long periods of time, threw everything into confusion, so we needed to replace all the ATC devices and improve the structures of the overhead contact wires.

Not all of these kinds of comprehensive replacement of facilities could be carried out during the brief nighttime maintenance period. An emergency measure directed all *Shinkansen* trains to stop running for half a day at a time several times a year so that crews could replace and repair the tracks and other facilities.

Nobody had thought that the *Shinkansen* facilities would start causing so much trouble so soon because things were completely different on conventional lines. Yet this was undeniably one of the worst periods for labor relations in the history of JNR, and it was difficult to ensure that workers would maintain, inspect, and work on the equipment fully.

In any case, thanks to their experience in dealing with all these troubles, JNR not only made improvements to every area but also acquired the know-how to maintain, inspect, and modify the facilities.

To give just one example, the style of the train with a round nose sticking out in front, the train that shaped public perceptions of the *Shinkansen*, was called the 'the Series 0 types EMU' within the company, and 360 of these cars were manufactured and ready before the line opened. More and more were manufactured, a total of 3,216 cars by 1985. The first rolling stock manufactured was called 'first generation,' and the number of the 'generations' was increased to 'second generation' or 'third generation,' for example, each time improvements or modifications were made since

Fig. 5.7 Series 0 EMU running at high speed. *Photo* provided by Katsuji Iwasa

these improvements required changing the overall design. The Series 0-type EMUs are now in their 38th generation (Fig. 5.7).

The *Shinkansen* has experienced a lot of problems, including breakdowns of equipment. Most of them have impeded correct and orderly operation, but few of them have had any bearing on safety. Yet we have experienced some incidents, however few, that had to do with basic safety.

The first occurred on April 25, 1966, about a year-and-a-half after the beginning of operations, soon after the *Hikari 42* left Nagoya. The conductor in the last car noticed some sort of unusual swaying, and as the train approached a curve near Toyohashi Station, he not only heard an unusual noise but saw sparks flying. He quickly contacted the driver, who stopped the train.

An investigation revealed that an axle on the last car had broken. It evidently broke shortly after leaving Nagoya but was able to keep running fairly well and stay on track with the axle supported by the gearbox. We were very lucky that time.

After that incident, we enhanced our inspection routines by using ultrasound to look for damage to the axles. Since instituting these painstaking inspections, we haven't had any more incidents with broken axles, but the prospect is frightening.

The next incident happened on November 12, 1967, when the *Kodama 271* overshot Gifu-Hashima Station. The *Kodama* trains that served Gifu-Hashima were naturally expected to stop automatically with the ATC brake system. In the unlikely event that it passed by its expected stopping position, the train was equipped with an emergency brake. Despite that, the *Kodama* passed straight through Gifu-Hashima Station and didn't stop until it was on the bridge over the Nagara River.

This was an incredibly abnormal situation. An investigation showed that nothing was wrong with the ATC apparatus itself. Instead, the incident had been caused by slippage of the wheels.

The biggest accident, and probably the only derailment of a *Shinkansen* train, was the derailment of a deadheading train near Shin-Osaka Station on February 21, 1973.[7]

It was leaving a depot called the Osaka Operations Depot and heading toward Shin-Osaka when it ignored an ATC stop signal and continued forward onto the main track, where it absolutely should not enter (Fig. 5.8).

Since the turnout was open in the proper direction for mainline trains heading from Kyoto to Shin-Osaka, it broke. On top of that, the driver panicked and tried to reverse the train, which caused it to derail. If another train had been running on the main line at that time and place, there would have been an unthinkable disaster. However, the moment the deadhead train intruded onto the track, other trains on the main line were stopped by the ATC stop signal system.

This was a terrible accident. The ATC system was the apparatus that was supposed to stop the train in case of danger, even if the driver didn't do anything. An accident

---

[7] The Joetsu Shinkansen *Toki 325* train derailed on October 23, 2004, due to the Chuetsu Earthquake in Niigata Prefecture, and a Tohoku Shinkansen train derailed on March 16, 2022, due to the Fukushima Prefecture Offshore Earthquake. In both cases, emergency brakes were activated by the Euredas, early earthquake detection and warning system, and no one was killed or injured in the derailments.

Fig. 5.8 Railway sleepers broken when a deadheading train failed to stop at an ATC stop signal and derailed. February 21, 1973, near Shin-Osaka Station. *Photo provided by Tokyo Shinbunsha*

that wasn't supposed to happen had indeed happened to one of the essential safety devices.

An investigative team looking for the cause of the accident concluded that the ATC brake had actually worked but that the wheels had slipped due to oil having been applied to the tracks to reduce friction. This had caused the deadhead train to overshoot the stop signal. In addition, the distance available for stopping was quite short at this particular place. After this accident, a variety of measures were implemented, including requiring trains to stop before stop signals and improving the ATC system.

JNR experienced one more unusual accident with the ATC system. On September 12, 1974, a Tokyo-bound *Kodama 120* was approaching a depot near Shinagawa when there was an indication from the ATC stop signal. But just before the train stopped, the signal changed to another indication that meant it was all right to proceed forward at 30 km/h. Yet when the driver looked ahead, he saw that the turnout was open not toward the main line but in the proper direction for trains coming out of the depot onto the main line. He stopped the train and reported the malfunction to the Control Center. This, too, was something that supposedly couldn't happen with the ATC, and it was something that must never happen.

The cause of the malfunction is too technical to be understood by anyone but experts, but there was no doubt that something was wrong with the ATC. In response, the ATC system underwent further improvements.

I think that the birth of the *Shinkansen* and the subsequent struggles with malfunctions have taught us something about the essence of systems and technology.

First of all, innovations occur in basic technology in various fields. In that process, many attempts succeed and fail, and after tests and improvements, we end up with a usable product. Television grew out of the invention of the transistor and the cathode ray tube, while the *Shinkansen* grew out of innovations in rolling stock, track, signals, and power technology. When several types of technologies are available and brought together to produce a new system or merchandise concept, the result is a hit product. Hit products arise out of innovations in basic technology and new concepts, and the *Shinkansen* was JNR's hit product.

However, new products frequently meet obstacles, namely initial breakdowns and unforeseen problems. In the world of computers, these are called 'bugs,' and experts are aware that every new product has bugs. No matter how seriously and intently you discuss and investigate a question, no matter how many tests you run, it is nearly impossible to find and eliminate all the bugs ahead of time. We won't find them unless we actually use the product.

The *Shinkansen* also had a great many bugs, big ones, and small ones. It took until 10 years after the opening of the line for all the bugs to be eliminated. However, we cannot guarantee that there are no bugs left, and new bugs may show up at any time. We don't believe that the *Shinkansen* is a perfect system, and we don't believe in the 'myth of safety.' The important thing is to be constantly on the lookout for bugs and eliminate the small ones before they turn into big ones. Nothing that human beings make is perfect. However, there is no doubt that the *Shinkansen* is the safest of all the transportation systems that human beings have created.

5.10   The Groundbreaking Mutual Through Operation Plan

I think it was in 1971 that I found an article like the title of this section in the influential French newspaper *Le Figaro*. It said "In Japan, JNR and other railways and subways that extend into Tokyo's suburbs are offering through-operation service on one another's tracks, so it is very convenient for the riders. France ought to introduce a Japanese-style mutual through operation plan, too."

As you know, large European cities such as London and Paris have huge terminal stations, but for a long time, they didn't have tracks extending into the city. Paris and large cities in Germany now have Japanese-style mutual through operations, but at that time, they did not yet exist. So the *Shinkansen* is not the only railway concept that Japan invented. The existence of so many private railways and their success in diversified management may also be a new model for railways, at least from a management point of view.

The first time that a suburban railway and subway set up mutual through operations was in 1960 when the Keisei Electric Railway and Toei Rapid Transit Authority's Asakusa Line built a mutual through operation plan. The first mutual through operation plan involving JNR and a subway took the form of those on JNR's Chuo Line and the Tozai subway line.

I was put in charge of these first mutual through operations, having become Head of the EMU Division, Train Operations Department, Tokyo Railway Administration Bureau. It was an overwhelming position, responsible for all the transport plans for EMU trains in Tokyo, all the operation plans of rolling stock and crews, all inspections and improvements, and the education and training of drivers. Frankly, it was a heavy burden to place on a 32-year-old like myself.

The part of the job that was the biggest ordeal for me was handling labor problems. The Tokyo EMU Department was the most combative section of the JNR Workers' Labor Union. Whatever we tried to do—revising the schedule, introducing new rolling stock, transferring to a new depot—we had to suffer through extremely annoying labor union-management bargaining. Once I confronted the union members alone for over six hours without a break, unable even to go to the toilet, surrounded by well over 200 shouting, jeering people. Since I would lose if I didn't say anything, I kept talking. After the bargaining was over, I raced to the toilet, but I was so tense that my physiology did not work well for a while. That's the only time in my life that such a thing ever happened to me.

In those days, the chairman of the Tokyo Workers' Labor Union was Mitsuo Tomizuka, who later became Secretary-General of the General Council of Trade Unions of Japan. Looking back, I can't think of any other time in my life that was so psychologically stressful. Sometimes I even wished that there'd be an earthquake so that I wouldn't have to attend the next day's labor union-management bargaining. In addition, this was a period marked by one project after another.

In just two years, the number of EMUs in Tokyo increased from 3,700 to 4,100, and new depots, the Shinagawa EMU Depot and the Toyota EMU Depot, were completed. The double tracks on the Chuo Line between Takao and Kofu was completed, the switchbacks were eliminated, and new stations appeared on steep gradients at Hatsukari and Sasago. The *Azusa* limited express between Shinjuku and Matsumoto and the *Asama* limited express on the Shinetsu Line made their first runs. The elevated structure of the quadruple tracks on the Chuo Line between Nakano and Mitaka was completed, and the first special commuter express EMU trains ran on the line, offering mutual through operation service with the Tozai subway line.

The *Shinkansen* was certainly important, but relieving the congestion on Tokyo's EMU commuter trains was also an important issue. The commuter EMU trains were much more overcrowded than they are now, and the tactic of alleviating congestion by increasing the number of trains had reached its limit. The time had come to build new tracks.

In 1965, JNR launched its 'Five Direction Strategies.' The 'five directions' were the Tokaido, Chuo, Tohoku-Takasaki, Joban, and Sobu Lines, and the idea was to build new double-track lines in their respective directions. It was a big project, with a price tag of 480 billion yen.

It's hard to believe now, but in those days, EMU trains of the Tokaido Line and the Yokosuka Line ran on the same tracks, and the EMU trains of the Tohoku-Takasaki Line used the tracks of the Keihin Tohoku Line. There was no distinction between the express lines and the local lines in the Chuo Line, the Joban Line, and the Sobu Line.

Building new tracks in a densely populated city would be a tremendous construction project, especially extending the tracks into the central area. There was no room for new platforms at Tokyo or Ueno Stations. The only solution we could find was to connect to subways for coming into the center of the city. That's when we hit upon the concept of connecting to the subway lines as well because we didn't have the kinds of enormous terminal stations found in European cities.

Mutual through operations are far from simple. What should the standard design of the rolling stock be? Which partner should make what? How would the fares be calculated? What about the train schedule? Another troublesome aspect was that JNR and the subways had different signal systems. JNR commuter trains had ATS, while subway cars had ATC, and the newer EMUs had to be built with both systems. For that reason, JNR EMUs operated onto the subway lines became the first of the non-*Shinkansen* trains to have ATC installed.

The train schedules did not present any particular problems. All the two companies had to do was sit down and discuss them, and there were rarely any serious disagreements. When subway EMUs ran on JNR tracks, the income from that section was considered JNR income, so JNR had to pay rolling stock usage fees to the subway company. A similar arrangement in the opposite direction prevailed when JNR EMU trains ran on subway tracks.

Since that was a lot of bother, the principle was established that the number of kilometers that JNR cars ran on subway tracks and the number of kilometers that subway cars ran on JNR tracks should end up being equal so that no detailed calculations were needed. However, there were times when the distances were not the same, as well as imbalances caused by schedule disruptions or trains breaking down. At such times, they would make adjustments by paying rolling stock usage fees. Tickets and commuter pass for mutual through operation service would be calculated one by one, based on the segment the passenger had ridden. Then, just before the shared service was about to begin, we were faced with problems concerning inspections by the Ministry of Transport.

Since JNR was a government organization, it could inspect its own rolling stock and facilities as they were completed, but private railways, including subways, had to be inspected by the Ministry of Transport. This meant that rolling stock of JNR that would run on the subway tracks would also have to undergo this inspection. Even though I thought that this mere formality of an inspection was meaningless, I rode along on a test EMU train. When we made the signal at Takadanobaba Station of the subway turn red and tried to make the train run, the ATC brakes did not work for some reason, and the train passed right through the red light. I was shocked and horrified.

Of course, JNR's own inspections would have found this malfunction. It was just ironic that it had occurred during the inspection by the Ministry of Transport, and it was a strange twist of fate that only the test EMU train seemed to suffer from this abnormality. When I thought long and hard about what had happened, I realized that since there had been no ATC sections within the JNR commuter system, we had done hardly any pre-service test runs using ATC. I supposed this was one of the pitfalls of bringing about a new transportation system like the mutual through operation system.

Later, in 1971, the Joban Line and the Chiyoda subway line began mutual through operations. At present, there are seven such mutual through operation arrangements with subway lines and suburban commuter lines, including some private railways. In addition, the JR Yokosuka and Sobu Lines may be regarded as a kind of mutual through operation with a subway.

Europe has also adopted this Japanese idea and instituted mutual through operation service between national railways and subways. In these European-style arrangements, such as the express subway line (RER) in Paris or the S-Bahn in Frankfurt or Munich, not only do the trains run straight through on tracks belonging to another railway, but the different lines have a uniform fare system.

One way in which Europe differs greatly from Japan is that its urban railways everywhere are losing money and require a lot of funding subsidies from the central and local governments. It is not unusual to find cities where the government subsidies are larger than the income from ticket sales. It makes me wonder whether the lack of entrance gates one often sees in Europe's urban transportation systems is due to the confidence that the government will provide subsidies, even if the system is in the red.

Sometimes special ticket inspectors make the rounds, and when they find someone riding illegally, they levy a fine equal to about 20 times the fare. While riding the S-Bahn between Frankfurt and its airports, I witnessed a scene in which three burly inspectors came around and reproached a passenger who looked like a foreign laborer for riding illegally. They were so stern that they were scary. By world standards, railways in Japan's major cities are exceptional in being managerially independent and profitable. They are a precious institution.

Chapter 6
Safety and Comfort

Under the author's leadership, safety has dramatically improved since the company can now afford to invest more in safety after becoming JR. This was based on his experience working for JNR during the period when the *Shinkansen* was in operation, when he was forced to deal with the train accidents that occurred amid labor-management confrontations, and his experience working at UIC in France, where he came into contact with people with completely different ideas and railway systems.

Furthermore, with regard to riding comfort, which is an important output of railways, the author experienced how the creation of a comfortable train can be done only when knowing how passengers feel about it rather than simply making judgments based on data.

This chapter provides material for readers to realize that high-quality railways, which now have 30 seconds of delay time per train, are not made overnight and that the world's railways can become higher-quality ones.

6.1 The Battle against Accidents

In 1968, I was promoted from Head of the EMU Division of the Tokyo Railway Administration Bureau to the Safety Division at the JNR Headquarters. My official title was to be Assistant Manager of the Safety Division, Train Operations Department, and my new office was in charge of investigating accidents, developing safety measures, and enacting operation regulations.

I have to relate one particularly humorous story that happened soon after I was transferred.

One day, my wife came home from my son's kindergarten laughing. She had explained the transfer to my son's teacher, who had smiled and replied, "That figures."

At the Tokyo Railway Administration Bureau, I was so exhausted after work that I had not been at all involved in the lives of my children. I had never even been to their

© Japan Railway Technical Service 2024
S. Yamanouchi, *If there were no Shinkansen*,
https://doi.org/10.1007/978-981-99-8890-7_6

kindergarten. Knowing that I was a terrible father, the kindergarten teacher apparently assumed that, even though I would now be working at the JNR Headquarters, I had been demoted from Manager to Assistant Manager.

What was waiting for me after the hell of the Tokyo office entailed much more responsibility than is normally given to someone of my age. However, it was not paradise. I had been condemned instead to purgatory.

Train accidents were a frequent occurrence during the year that I was promoted to the Safety Division. Trains were involved in a serious accident nearly every month, and how to respond and what measures to take were the bane of my working life then. The records from 1968 show just how bad the situation was.

Jan 18

A freight train derails after proceeding against a red signal from the Shikagoe Signal Station on the Nemuro Main Line in Hokkaido; service suspended for 12 hours.

Feb 2

A dozing driver of a locomotive derails 14 freight cars when pulling into Higashi Shikagoe Station on the Nemuro Main Line in Hokkaido; service suspended for more than six hours.

Feb 15

A passenger train waiting for the go ahead at Maibara Station departs on the green signal from the freight line. The driver mistakenly believes he is running on the passenger line and obeys the passenger line signals along the track running parallel to the freight line the train is actually running on. A passenger train collides with the freight train from behind; the passenger train derails and overturns, injuring five people.

Apr 8

A DMU train and dump truck collide on the Geibi Line; the train derails, and 22 people are injured.

Apr 19

An EMU train collides with a freight train from behind on the Joetsu Line; both trains derail, and 69 people are injured.

May 16

The Tohoku Main Line where double tracking was underway sustains significant damage from the Tokachioki Earthquake, a 7.8 magnitude earthquake on the scale of the Great Kanto Earthquake. Service at the Seikan Ferry port facilities is suspended for five days. A locomotive and 13 freight cars derail.

May 17

Excessive speed when entering the passing loop on the Kagoshima Main Line at Togo Station causes 26 freight cars to derail and overturn at the turnout.

Jun 16

The explosion of a timed bomb in a bag placed on an overhead rack injures 29 passengers at Ofuna Station on the Yokosuka Line.

Jun 27

Excessive speed causes the locomotive and 30 cars of a freight train to derail and over-turn when passing through Zeze Station on the Tokaido Line. Service is suspended for 28 hours. Excessive speed was due to a dozing driver.

Jul 16

An EMU train collides with the rear end of an EMU train stopped near Ochanomizu Station on the Chuo Line, and 210 people are injured. The EMU train had pulled out of the station but then made an emergency stop to deal with a passenger's hand caught in the door. This train was rear-ended by the following train. Despite a red home signal, the driver of the second train assumed that the preceding train was already clear. The driver in charge disappeared temporarily from the site after the accident, prompting extensive newspaper coverage featuring headlines like "Big Train Accident Caused by Hit and Run Driver."

Sep 21

A DMU and dump truck collide at Tazawako Station; three cars derail and 13 people are injured.

Sep 24

A *Shinkansen* train hits crews preparing for work on the Tokaido *Shinkansen*; three workers are killed and nine are injured.

Sep 27

A DMU and minibus collide on the Taisha Line, injuring 21 people.

Sep 29

The driver and assistant driver of a locomotive, both drunk on whiskey, fought to stop the freight train that they are operating in the middle of the Oume Line to fight.

Oct 1

A bridge on the Furano Line collapses, plunging a freight train into a river. Three people, including the driver, are killed.

Oct 12

An express train scheduled to stop at Naie Station on the Hakodate main line fails to stop and passes through the station. The crossing barrier at the level crossing remains up, and the train collides with a dump truck, derailing four cars and injuring 29 people.

Nov 16

An EMU train and dump truck collide at Aboshi Station on the Sanyo Line, injuring 12 people.

Furthermore, on February 13, 1969, a total of eight railway workers, some of whom were doing track maintenance at the time, were hit and killed by EMU trains in two separate accidents on the Hakubi and Sakurai Lines. This is extremely unusual, and it was hard to believe that two of these accidents occurring at the same time in two different places could be a coincidence. It seemed to be a warning from God.

All of these accidents, with the exception of the last two, occurred within less than a year. All of our time was taken up by dealing with accident after accident, investigating the causes, and trying to understand what had happened. I often ended up sleeping at the office. Top executives were angry with me; I was hounded by the media. I would be called into the supervisory authorities who would scold me and demand detailed reports of the accidents and measures to prevent future accidents. I was even ordered on occasion to appear before the police. To be honest, there was little I could do to resolve these problems.

In October 1968, JNR timetable had been revised on an unprecedented scale. October 1968 corresponds to October of Showa 43 in the Japanese calendar, so these revisions were nicknamed 'four-three-ten.' Work of electrification and double tracking on the Tohoku Main Line had been completed, and the number of limited express trains dramatically increased throughout Japan. This dramatic expansion of work volume was, perhaps, not unrelated to the number of accidents occurring around this time. Detailed descriptions of each accident, however, show a striking number of basic mistakes by drivers and level crossing accidents, and it is here that the essence of our problems lay.

Around this time, the conflict between JNR labor unions and management intensified, which exacerbated discipline in the field and weakened our instruction system. The company's financial situation, which had ironically showed its first deficit just as the Tokaido *Shinkansen* service was introduced, was also deteriorating dramatically. In contrast to the enormous capital investments that were being made, passenger and freight revenue had not increased by as much as expected. The only exceptions were the *Shinkansen* lines.

Japan had entered the era of motorization, and the rapid increase in the number of private cars was having a particularly serious effect on freight transportation.

6.2 Intense Labor Union Opposition

It was at this time that JNR set about rationalizing its operating system, which drew intense opposition from the labor unions. JNR planned to cut onboard driving crew from two motormen to one in 1969, but these efforts generated particularly intense conflicts between labor unions and management. Although at least two crews, the locomotive engine driver and the assistant driver, are required to operate a steam locomotive, there is no need for an assistant on electric and diesel locomotives. A single driver in fact, originally operated the EMUs.

Regardless, the labor unions fiercely opposed halving the number of driving crews, claiming that safety would suffer as a result.

Management rejected the idea that two driving crews necessarily ensured safer train operation. We countered that safety concerns would be mitigated by safety devices that sound an alarm when no activity has taken place in the locomotive for one minute or longer and automatically stop the train when there is no response to this alarm.

This device had been used widely on European railways for some time, but strong union opposition had made it difficult to install these devices on Japanese trains.

The conflict between labor unions and management spilled onto the worksites, and field discipline quickly deteriorated. On some worksites, it had become a place of conflict over daily work more than instruction and training.

A lack of order and morale among employees working on train operations, as well as inferior safety equipment, were, I believe, the essential reasons for the large number of accidents seen at this time. The Automatic Train Stop (ATS) device that was then used was totally inadequate, and level crossing barriers and signals were quite primitive compared to our current equipment. These accidents frequently frustrated top executives, who would bellow, "What the hell are you people in the Train Operations Department doing? You must pay more attention to your jobs!" They would demand that new safety measures be drawn up. But it was not an essential solution. To top it all off, partial restructuring of the organization and executive shake-ups were also attempted. These were nothing more than band-aid solutions and bureaucratic responses to a real problem.

Although I no longer recall the particulars, I once reported on a major accident directly to Minister of Transport Yasuhiro Nakasone himself, who had arrived at the Shinjuku Station Master's office from Yamanashi Prefecture. After listening in silence, he replied simply, "JNR mid-level leaders are collapsed."

The assistant station managers and locomotive driving instructors were the ones at JNR who were truly responsible for safety in the field. He pointed out that this most important part of the organization had become impotent through the unproductive labor union-management conflict.

Order and discipline in the field, restoring motivation, and the improvement of ATS and other safety devices were the most compelling issues for improving safety during this period. It was nearly impossible, however, to do anything about these crucial issues. In reality, the reverse was actually occurring. The system of resolving issues in the work sites, which later paralyzed discipline in the work sites, was instituted that year in the form of a promise made by management to the labor unions. Even though we all knew how totally inadequate the ATS device was, the technology for a new automatic train stop device to replace it was not yet ready.

ATC devices, which had been installed on *Shinkansen* lines and was being introduced on a number of private railways, were technically incompatible with freight trains. An improvised, technically compatible ATC had been developed, but JNR's rapidly deteriorating financial situation made the enormous capital investment required to install this device on the 20,000 km of track length under JNR jurisdiction unacceptable. Huge amounts of capital investment were still needed to construct new

Shinkansen lines, increase the number of commuter EMU trains, double tracking, and electrify lines. The new ATS system and restored order in the field would both have to wait until JNR was privatized.

Despite all this, statistics show that there were fewer accidents that year than the year before. The term 'accident' may be defined in many ways; 'train accidents' refer to those in which a train collides with another one, derails or catches fire, and these present the greatest danger. The 83 train accidents in 1968, nine less than the previous year, indicate the low level of safety at the time.

In 1956, the year I joined the company, JNR saw 139 train accidents, which averages out to a train derailment or collision every three days. This was in spite of the fact that trains only ran half the total number of kilometers they do today. The largest number of train accidents in a one-year period, 742 accidents, occurred in 1944. This average of two train accidents a day occurred at the end of World War II, a highly unusual period for Japanese railways.

To put these figures in perspective, the majority of the 12 train accidents occurring in 1997 were minor ones involving the derailment of a railcar wheel due to rain or falling rocks. However, there were also some serious accidents during this year as well, such as an EMU train rear-end colliding with a freight train on the Tokaido Line in August and a collision at Otsuki Station on the Chuo Line in October. We have reduced train accidents substantially, but there are still safety issues that must be resolved (Fig. 6.1).

Fig. 6.1 Evening, October 12, 1997. *Super Azusa 13* limited express collides with a deadhead EMU train at Otsuki Station on the JR Chuo Line. Safety issues remain unresolved. *Photo* provided by Tokyo Shinbunsha

In Japan, we do not clearly distinguish between 'operating accidents' and 'accidents.' When train service is suspended because electricity is not being transmitted to EMU trains due to a breakage of the overhead contact wire, newspaper headlines read, "JR Stops Trains Due to Overhead Contact Wire Accident." In the U.S. or Europe, the term 'accident' does not refer to a trouble of this type; it is reserved for situations in which people are killed or injured or in which trains or other vehicles get damaged.

6.3   Operating Incidents

Troubles with stoppage of trains due to the EMU failure or the failure of the rail are called an 'incident' in the U.S. and Europe and are therefore clearly distinguished from an accident. During the 1970s, this concept was also studied in Japan, and regulations were enacted that clearly distinguished between 'operating accidents,' or what were termed 'accidents' in the West, and 'operating incidents' involving the suspension of operations but no direct harm to persons or structures.

After the privatization of JNR, operating accidents at East Japan Railway Company were reduced by more than half. Even more dramatically, level crossing accidents were reduced to less than one-third over ten years.[1] A major reason for this reduction was the introduction of a safety device called the level crossing obstacle detection sensor. With this device, laser beams identify when a car or other obstacle is blocking a railway level crossing and automatically activate an emergency stop signal that stops a train.

Although the technology for this device was nearly ready during JNR era, no real effort was made to develop it for installation. The device effectively prevents level crossing accidents but can be too sensitive, with emergency stop signals activated by heavy snow or at times by a dog running across a railway level crossing. A railway worker must then rush immediately to the site to deactivate the triggered emergency stop signal or it will disrupt train operations, which is why JNR was reluctant to install the devices. Aggressive steps after JNR became JR brought about a dramatic drop in level crossing accidents. This is one form of radical change to the corporate culture.

The largest number of level crossing accidents in a one-year period, 3,223 accidents, occurred in 1961. In 1997, the total number of level crossing accidents among all of the JR Group companies was 275. How significant the prevention of level crossing accidents was when we consider how radically the number of cars on the road has increased over that period. Ironically, it is this substantial reduction that has increased the number of operating incidents because a train is now stopped by the emergency stop signal before it collides with a car or other object that is blocking a

---

[1] As of 2021, the number of 'operating accidents' had been reduced to one-third and level crossing accidents to one-fifth, compared to the time of privatization.

level crossing. The basic policy of JNR was for any driver sensing danger to immediately bring the train to a stop, as well. The action itself of stopping a train before it
reaches a level crossing is not, therefore, a problem, but technologies must be refined
to further reduce railcar and signal device malfunctioning.

How does the safety of Japan's railways compare with railways overseas? I was
shocked to see data showing more than 7,000 accidents involving derailments on
U.S. railways in one year, but the freight-centered railway system in the U.S. does
not offer a proper comparison for the Japanese railway system. The railway situation
in Europe is comparable to that in Japan, although the idea of what constitutes an
accident is slightly different. In 1996, SNCF saw 78 accidents that would be defined
as 'train accidents' in Japan. Germany does not publicly release its official data, but
the number of accidents in that country apparently exceeds that in France.

6.4 Average 30–40 Seconds' Delay

The aspect of its railways that Japan can take the most pride in is the precision
of its train schedules. There are periods of instability like the successive troubles
on East Japan Railway Company's Chuo Line in 1998, which left us at a loss as
to how to justify them to the public. Surmounting these problems, however, brings
technological advances and improves safety, and the steady drop in accidents on the
Japanese railway system can be attributed to the persistent efforts in this area.

Some 13,000 trains operate daily on JR East lines. Train delays average about
40 seconds on *Shinkansen* lines and 30 seconds on meter gauge lines. This is the
level of schedule precision we see generally each year, and the average delay on the
Tohoku-Joetsu *Shinkansen*, which has rarely experienced problems, is actually only
10 seconds. This can also be attributed to the improvements made in each department
as a result of initial train failures and other problems on the Tokaido *Shinkansen*.
The Tohoku-Joetsu *Shinkansen* lines have recently seen a slight increase. It is now
16 years since the service was introduced on this line.[2] These problems may be the
result of the type of aging we saw in the tenth year of Tokaido *Shinkansen* service,
and it is perhaps time for the equipment to be replaced. A railway system is only
truly developed when operators have the knowledge required to predict when and
where railcars and other equipment will break down.

The mean time between failures (MTBF) is an index used to objectively measure
the reliability of a given system and refers to the average length of time between major
breakdowns. The concept apparently evolved in the U.S. when engineers working
on the Apollo mission needed a method of controlling the reliability of parts.

To experiment with this concept, I have calculated the MTBF figure for *Shinkansen*
operations.

---

[2] As of 2022, 30 years will have passed since the opening of the Tohoku and Joetsu *Shinkansen*
lines.

My hypothesis was that a problem causing a delay of one hour or longer would occur when a *Shinkansen* train had been operated for a certain number of hours. My calculations indicate that the MTBF figure for the Tohoku-Joetsu *Shinkansen* was approximately 30,000 hours in 1997. In other words, there is the possibility of some type of failure occurring when a train has been in operation for a total of 30,000 hours; this converts to the possibility of a major failure six times in a year. This method deals with the reliability of the entire system, not only the equipment, and the figure reflects delays caused by earthquakes, typhoons, and other natural disasters as well.

In January 1998, problems at an electricity substation, which had begun on New Year's morning, generated Tohoku-Joetsu *Shinkansen* failures that disrupted operations and resulted in the suspension of train operation service. My calculations indicate an approximately 5,000-hours' MTBF for these lines during January.

The repeated schedule delays on the Tokaido *Shinkansen* between 1974 and 1976 were severely criticized in the press with headlines reading, "*Shinkansen* Service Suspended Yet Again." The MTBF figure during this period was a paltry 4,000–8,000 hours.

What about conventional lines in terms of MTBF? In January 1998, there were also repeated schedule delays on the Chuo Line due to various causes including a fatal accident. While the Chuo Line's MTBF was 6,000 hours during the year 1997, this fell to 3,000 hours in this month. Trains running on the Chuo Line were frequently delayed throughout this entire one-year period.

These figures seem to indicate that,

> At its present level in Japan, the railway MTBF should reach 10,000 hours or more. If and when this figure reaches 30,000 hours, we will be able to lay claim to an 'exceedingly precise train schedule.' On the other hand, a deluge of complaints will follow a drop in MTBF of half or more on even a single line.

Railway experts visiting Japan from overseas are amazed by the cleanliness of the streets and the precision of the train schedule. Standing on one of the Tokyo Station platforms of the Tohoku-Joetsu *Shinkansen* and watching one train after another arrive at the only two tracks, the cars cleaned within a dozen minutes and the train pull out again. The expression on their faces says beyond admiration, "Japanese people are a different breed altogether." Putting aside for the moment the question of whether Japanese people are actually different, the scarcity of tracks and lack of facility space requires Japanese railway workers to have this 'gene.'

Two years ago, I came across an article in a French railway magazine with the shocking headline, "Approximately One in Six Trains Delayed." This headline seems to accurately describe the condition of the approximately 5,000 trains a day operated by SNCF in the Paris suburbs in 1995. In Japan, a train is recorded as delayed even if it is delayed by only a minute. On most *Shinkansen* and other train lines, an announcement apologizing for the delay is made when a train is running two or more minutes behind schedule.

In France, however, short-distance trains are recorded as being on time unless they are delayed by more than five minutes, with long-distance trains being considered delayed at fifteen minutes or more. Even with this leeway, one in six trains runs behind schedule.

The situation with DB was even worse. In 1995, 20% of limited express and express trains were delayed by five minutes or longer. DB began a concerted effort to reduce the number of delayed trains around 1997, since these delays affected the public's trust in their rail system.

In May 1997, a two-day symposium on precision train operations was held in Leipzig, and I was invited to give a talk and take part in the panel discussion. This symposium gave many workers at the forefront of train operations the opportunity to make suggestions and listen to a speech by the country's Minister of Transport. This symposium may have contributed to the considerable improvement in train schedule precision since then. Since 1998, 93% of all limited express and express trains have been operating within five minutes of their scheduled times.

On a recent trip to Germany, I noticed red lines drawn on the platforms at major train stations. Excessive passenger boarding times are a major cause of train delays, and the Japanese custom of lining up to wait for a train is not prevalent in Europe. These red lines have been drawn, as in Japan, to indicate the location of train doors for arriving trains so that passengers will line up in these areas while they wait. The magnitude of train delays is now also taken into consideration in calculating employee bonuses.

Delayed trains affect station work schedules and disrupt planned maintenance tasks; they may even affect safety. Safety stems from precise train operations.

6.5   Experience in the International Business World

The railway system is essentially a domestic industry, and particularly in Japan, is not involved in international business or multilateral issues.

This is not true, however, in Europe which incorporates the borders of many countries. The systems in this region operate a large number of international trains, which require coordination of train schedules and unification of railcars and facility regulations imperative. The reality is considerably complicated; each country maintains different electrification systems, and the number of separate signal systems alone totals 16. Track gauges in Spain, Portugal, and the former Soviet Union countries also differ from those in the rest of Europe. Another of the many issues Europe faces is that standard continental railcars cannot operate on railways in Britain with their smaller tunnels and bridges. These problems were generated by the autonomous development of railways as independent domestic systems.

Progress on these issues, however, began from the 1870s with the construction of major trunk lines that enabled trains to travel across borders. In 1883, the famous *Orient Express* was introduced.

The Conference on European Train Schedules (CEH), the first international European railway organization, was held in Cologne, Germany, in 1872. It was in this exact year that Japan began to operate its first trains. At that time, a number of railway companies operated within each European country, and conference participants included representatives from 40 German railway companies, as well as from railway companies in Austria, Belgium, the Netherlands, Romania, and elsewhere. French and Italian railway companies did not send representatives since they were unwilling to participate in any German-led international conference.

The German empire, which conquered France in 1870, was looking to control Europe through the railway system. As this conference was obviously part of its plan, it was natural that France, the neutral state of Switzerland and other countries chose not to attend.

The International Railway Congress Association (IRCA) was formed in Belgium in 1885. This association remains active today, holding an international congress every two years with railway company representatives from around the world. Members of this association comprise 94 railway companies from around the world, including Japan's JR Group. In 1997, at Marrakesh, Morocco the 27th International Railway Congress was held on the topic of 'Railways –from a Necessity in the Economy to Financial Independence.' Nearly 1,000 participants attended this four-day congress, the largest international railway community conference ever.

The neutral location of the IRCA Secretariat in Belgium prevents game playing and contests of pride between powerful countries. After World War I, the SBB was appointed Secretariat of the European Timetable Conference/Conférence Internationale des Horaires (CEH), and this conference was restructured in 1997 as the Forum Train Europe. The Forum retains its original role of coordinating the schedules of international trains.

Each of the associations mentioned above was formed as a conference organizer, with the national railways discharging their secretariat responsibilities in their spare time. The International Union of Railways (UIC) is the only fully functional international railway association with more than 100 full-time employees at its headquarters in the 15th District in Paris. Established in 1922, soon after the end of World War I, this association primarily focuses on railway traffic policy, the coordination of passenger and freight transportation, technical standards, and technological development.

France founded the UIC after its victory in World War I to promote its leadership of European railways. It remains, however, the sole international cooperation organization in Europe to deal with the practical aspects of railway operation.

The strong overtones of French control, however, have kept the CEH from being affiliated with the UIC. Japan has been a supporting member since the association was founded. The reason for this, of course, is not that Japan was operating international trains, but because it was already selling tickets for trains to Europe via Siberia at this time.

In 1961, Louis Armand, the former president of SNCF was appointed Secretary-General of the UIC and set about reforming the association on a large scale. After his tenure as president of SNCF, Armand served as president of the European Atomic Energy Group (EURATOM) and was a passionate advocate of European unification. The French can be broadly divided into nationalists who emphasize the 'glory of France' and 'Europeanists' who consider Europe to be a single group and advocate European unification. Armand sided passionately with the latter group.

The successive generations of SNCF presidents appointed as UIC chairmen were replaced by a system under which the president of each national railway in Europe was appointed UIC chairman in turn. In the Secretariat, experts from all over Europe loosened the hold of French executives on this office. With these reforms, the UIC was reborn as a truly international association.

Despite the changes, however, this association still existed primarily to serve European railways. Japan and other Asian and African countries participated as supporting members, but they did nothing more than attend the general meeting or symposium held annually.

Out of the blue, Japan was asked to appoint one executive member in 1969, and for whatever reason, I was chosen to represent my country. A JNR executive had apparently broached this subject with Louis Armand at the Cybernetics conference in Paris the previous year, and he had acquiesced. This, I suppose, is what they mean by a bolt from the blue. When I returned home to tell my wife, once the position was formally offered, that I would now be working in Paris, she simply stared at me in mute amazement.

For one brief moment, I could only think that I was escaping the 'accident hell' of the Safety Division at long last. Then came the onslaught of anxiety.

First and foremost, I was worried about the language barrier. I had selected to take English and German for my high school and college language requirements. For the past three years, I had been studying French at home and at a language school, but only in my spare time. I would, of course, be the only Japanese person working there; I was the first non-European to serve as an executive of this organization. From the moment I stepped off the plane in Paris, I would naturally be speaking only French at work and around town.

I also worried quite a bit about the work that I would be expected to do. The group I was to be a member of was working on international cooperation and researching the use of computers in documentation. The general manager of SNCF Documentation Department served as group leader in between his other duties, and every member other than myself came from Germany and Romania. All of my colleagues would be fluent in French. A considerable amount of railway literature has been published throughout the world, and railways in each country have established their own libraries and cultural centers to catalog these publications. The publication of important materials in other countries often goes unnoticed elsewhere. Moreover, each country has developed its own independent data referencing systems. This setup is extremely inefficient in terms of documentation and referencing.

To improve this situation, my group was to establish a global document database at UIC Headquarters and develop a system under which computer terminals all over the world could reference this data.

Since I was not an information systems expert, my last days in Japan were spent rushing back and forth between the Railway Technical Research Institute, the Electrical Communications Research Institute at Denden Kosha (now NTT), and the Science and Technology Information Center to learn about documentation referencing systems from these experts.

The language barrier was, of course, the first problem I faced after arriving in Paris. I could manage my daily life but could not understand even half of what was being discussed at our group meetings. While I was busy trying to properly phrase what I wanted to say, the other members had already moved onto the next topic. I was in Paris for nearly six months before I was finally able to relax a bit during our meetings.

The next problem was finding a place to live. I studied the classifieds to find a suitable situation, but these advertisements seemed to be written in code. At last, I found a property that I was looking for, but I couldn't just drop by the real estate office as I would in Japan. In France, you must prearrange a day and time to meet over the telephone.

I finally met with the real estate agent to look at the available place but was told that I must call again to arrange another meeting to see the apartment. I faced a series of unpleasant *rendezvous*, the French word for appointment.

Finally, I found a three-bedroom apartment on my seventh or eighth *rendezvous* that seemed suitable. This furnished apartment was equipped with everything that I would need, from the dishes to the forks and knives. Strangely enough, I was asked to supply my own pillow, though.

The rental contract described everything in detail—the condition of the apartment, a list of the furniture and utensils, every scratch and dent—and stipulated that I would have to pay for any damage that I had caused when I moved out.

I had never had this experience in Japan. I was worried that there would be trouble later. The landlord was nice, and my anxiety proved unfounded, but even these trivial matters were confusing in a foreign culture.

My next struggle—a distinct feeling of isolation—surfaced after I had settled in and was working full-time again. After three months, I become somewhat accustomed to the French language and attempted to participate in our meetings, but the others paid little attention to me. I do not think that this was because my comments were meaningless. I may have been overly sensitive, but it seemed to me that my European colleagues were thinking, "What is this Asian kid going on about now?".

Lunch was yet another struggle with isolation. The employee cafeteria served hors d'oeuvres, meat dishes (an egg dish on Friday), dessert, coffee, and a small bottle of wine. No matter how fast I ate, this meal would take an hour, and our lunches were spent gossiping about various topics—politics, promotions, and the troubles of other employees. My colleagues listened politely the first two or three times that I spoke about Japan, but showed no interest whatsoever after this. I was careful until the end how to create a topic and enter into conversations.

The way in which Europeans worked was often confusing as well. Signatures and the minutes taken at meetings take on a special importance in a country that does not use name seals on official documents. The authorization of the minutes of important meetings would prompt several rounds of letters between officials. I was most surprised, however, by the letter our group received from a colleague working in the office next to ours.

The custom at UIC Headquarters at that time was that an attendant delivered documents and materials from other officials hourly.

Fig. 6.2 Author (right) taking part in an UIC (International Union of Railways) meeting. *Photo* provided by his family

I found a letter in the arrivals' box from Rafaju, an executive in charge of railcars who was a colleague in the next room (Fig. 6.2).

His letter read,

I would be honored to ask you a question I have regarding the following points. Regarding the regulation of international standard maritime containers, …

…Please accept my great respect and affection as I await your reply.

I rushed to find the data he requested and replied in kind, "I am honored to furnish the following reply to your inquiry regarding maritime containers."

The personal office culture in Europe was another surprise. Each UIC executive received his or her private office and secretary, as well as a pass entitling him or her to a first-class seat on all SNCF lines free of charge.

By contrast, the large, shared office spaces at Japanese companies apparently seem unusual to the French. A business trip report expresses the shock of an SNCF executive who visited Japan on one of the one-month exchanges with JNR held at that time. He writes, "I had heard about the lack of space in Japan that results from so many people living in such a small country. It was worse than I had expected. Even division managers share office space with their subordinates…".

When I had settled in somewhat with the European ways of doing business, I set about working on the concept of computerized documentation referencing system. I called for proposals to Bull Co., Ltd., a French company, ICL, a British company, and so on, and even went into the Nice suburbs to visit the IBM laboratory.

In the end, we settled on the GOLEM system developed by the German company Siemens, but I got the feeling that this decision was prompted by the political motivation of trying to incorporate DB.

The Siemens technicians were exceedingly confident that their system would enable us to immediately retrieve whatever information we were looking for.

In order to download this information from other computers, however, specialized software called a 'Thesaurus' would have to be installed in our system. The technicians persisted, assuring us that their computer had the capabilities to do whatever our system required. I found myself in a heated debate with these specialists over whether or not we actually needed this software, but finally persuaded him. Our group then developed a limited-scale 'Railway Thesaurus' software package for test purposes. We conducted a test run in the autumn of 1971 by accessing our database computer in Cologne from a distant computer in Munich and using a keyword to locate a required document.

The test was successful, and our group finalized our report on the concept of the system and test results, and they were submitted. Then our team was disbanded, and I returned home to Japan.

Our international documentation referencing system was never implemented. Since our most powerful proponent, former UIC Secretary-General Louis Armand passed away. His successor, Bemard de Fontgalland, was a pragmatist whose views differed greatly from Armand the Romantic. Armand had attempted too much. The UIC found itself in a precarious financial situation, and the majority of the projects that Armand had initiated were abandoned.

This experience in the world of international business taught me how difficult and time-consuming it is for more than one country to reach a consensus. France and Germany are at cross-purposes, particularly often. Germany has distanced itself from the strongly French-influenced UIC since it was founded and half-heartedly agreed to the concept of an international railway information center. In Germany, information is thought to be the possession of the nation, and this country would rather develop an information system encompassing all of its domestic industries than focus on the international integration of railway sector data.

Although the UIC is thought of as an international organization, its primary focus is, in reality, European railways. The first to voice opposition to this was Indian Railways (IR). India complained that the lack of benefits for non-European railways as UIC members was ludicrous in this day and age of internationalization. Even if the railway systems did not physically cross national borders, they argued, a number of global issues, such as traffic policies, railway privatization, data exchange, technical transfers, and stronger cooperation with manufacturers, needed to be addressed. Japan also stood in strong support of this perfectly reasonable stance.

6.6   Chairman of the World Executive Council

The UIC mobilized itself in 1995 and established the World Executive Council. The question of whom to appoint as its chairman, however, presented a problem. DB nominated Japan. The Director-General of UIC, who was at that time French, Walrave, was apparently opposed to a chairman from Japan. A subordinate executive came to my hotel room the night before the nomination was to take place. He asked me who I thought should be appointed chairman of the World Executive Council.

> I think it should be given to Musuva, the president of the Kenya Railways. He has a lot of energy and experience in international business.

He was obviously pleased by my answer and asked me to submit the nomination. I gladly accepted, and Musuva was selected as chairman the following day.

In 1996, however, political turmoil forced Musuva from his position as president of Kenya Railways. Council regulations stipulate that the World Executive Council Chairman must hold the highest position at a railway company.

In May 1997, I ran into Roumeguère, the new Director-General of the UIC, at a hotel in Stockholm the day before the World Executive Council meeting was to be held. I asked him.

> who would be selected as chairman.

Roumeguère was less cautious than his predecessor and replied straightforwardly,

> You.

> You´re kidding. The chairman should be selected from a developing country, don't you think? You don't really believe Japan should be selected, do you?

> No, I don't, but there is no one else to do the job. If you know of another qualified candidate, please tell me.

> India.

> The Chairman of IR holds office for only one year, so that's not possible.

> What about China?.

> They're not yet ready.

> South Africa is a possibility.

> The country still has the stain of apartheid.

U.S. railway companies and Russian Railways were not possibilities since they were not members of the UIC. These two powers are not concerned with French-led international organizations. Even so, the U.S. Federal Railway Administration (FRA) and the American Association of Railways (AAR) participate as supporting members, but the vestiges of the East–West conflict have kept Russia from participating.[3]

In opposition to the UIC, the Soviet Union established a separate international organization, the OSJD, during the Cold War. This organization comprised member

---

[3] Russian Railways has been a UIC member since 2006.

railways from Eastern European countries, Mongolia, North Korea, and elsewhere, and remains intact today. Eastern European railways are now members of both the UIC and OSJD. The OSJD meetings were apparently quite bureaucratic in style, and the following joke was often told when I was at the UIC. "The first four days of a five-day OSJD meeting are spent going over the minutes from the last meeting, and the fifth day is spent choosing the topics for the coming one."

Unable to suggest the U.S. and Russia, I found it difficult to come up with another candidate country for Roumeguère.

In desperation, I suggested another African country, Morocco.

This was flatly rejected, and asking the chairman to let me sleep on it, I returned to my room.

Larsson, who was both the Swedish National Railway (SJ) president and the UIC chairman, and Roumeguère, Secretary-General of the UIC, were waiting when I walked into the meeting the next morning. They asked for an answer.

I could only reply, "I thought about it all last night, and I cannot find a reason to refuse."

I understand. Coming from a Japanese person, that means yes.

Despite his grandiose title, the chairman of the World Executive Council does not hold a particularly powerful position. To be honest, I find regularly hosting meetings and seminars where not a word of Japanese is spoken to be quite bothersome. I am proud, however, of the unanimous election of a Japanese as chairman, and the Council Chairman is automatically appointed as UIC Vice Chairman. Japan has the *Shinkansen* to thank for these honors.

6.7 'Waltzing Train' Runs on the Chuo Line

In June 1972, soon after I had returned to Japan, I was appointed the Head of the Train Operations Department at the Nagoya Railway Administrative Bureau. The situation at this point was not just one of dreadful labor union-management relations; JNR itself was in terrible disarray. Efforts to improve productivity and their subsequent failure had further aggravated the conflict between labor unions and management, and unions were in complete control of the national railway. No decision could be made without the consent of the labor unions.

Take, as an example, a freight train that arrives at a station behind schedule. The labor union at worksites would argue; "This delay meant a change in the scheduled working hours that had been negotiated by the unions and management on the timetable revision. In their view, a change in working hours constituted a change in labor conditions. Workers would refuse to work until the unions held discussions on these new conditions."

While union representatives were standing around arguing their case, the train would pull out of the station again, still carrying the cargo that should have been unloaded. With repeat performances of these stand-offs, mountains of cargo accumulated on station platforms. Fruit and other fresh foods began to rot. When management tried to force employees to work, station masters and assistant masters would be tormented in kangaroo courts, and union-management negotiations at the Railway Administration Bureau would come to a halt.

I found myself the target of violence. Immediately after a workers' strike had ended, I had gone to check on the situation; at the conductors' office, I was surrounded by a crowd of workers who began jabbing and kicking me. Both the chief and the assistant station masters pretended not to notice the violence. Suddenly, someone grabbed me by the arms, threw me outside the office, and slammed the door shut.

I was later told that the leader of the local union chapter threw me out.

Despite the situation, orders from Headquarters to revise the train schedule and rationalize operations forced us to negotiate with the labor union in an effort to reach some sort of agreement.

In July 1973, electrification of the Nagoya—Nagano section of the Chuo Line was complete, and the *Shinano* limited express, the first tilting EMU train in Japan, was introduced. The tilting EMU train leans toward the outside of the tracks like a pendulum as the train enters a curve, enabling them to operate at a higher speed around the curve.

For railways in countries like Japan, which by necessity has many curved sections built into the rail network, increasing the maximum operating speed is extremely difficult. Mountainous sections like those on the Chuo Line are particularly winding. Rather than trying to increase the maximum train speed, the decisive factor in increasing the operating speed is to be able to take these curves faster.

The most effective way to do this is to set the outside rail higher than the inside on curved tracks. Railway terminology refers to this as 'cant.' The higher the cant, the faster trains are able to take curves. Roller coasters have a steep cant for incredible speed. Railways, however, also have to deal with slower trains, and trains must be able to slow for red and caution signals, which limits the possibility of an excessively steep cant. Another concern is strong crosswinds that may push a train over the inside rail on curves.

The tilting EMU concept came from efforts to allow trains to operate on curved sections of track without the cant being too high at a higher speed. Passengers in the car will feel their bodies being pushed strongly outward when EMU trains take low-cant curves at high speeds. This is called 'centrifugal force.' Being repeatedly subjected to a strong centrifugal force makes passengers uncomfortable. The tilting EMU train offers a solution to this problem; it mitigates the impact of the centrifugal force on passengers by tilting the car bodies themselves toward the outside rail as the EMU train approaches a curve.

The maximum operating speed of the *Shinano* limited express is 120 km/h, and this train can take curves about 10 km/h faster than a normal limited express train. The *Shinano,* however, only reaches 120 km/h in three minutes on a section between Nagoya and Nakatsugawa, which illustrates the absurdity of trying to increase the maximum operating speed on this type of track (Fig. 6.3).

Fig. 6.3 1973 First tilting EMU train, the *Shinano* limited express. *Photo* provided by the Railway Museum

The concept of the tilting EMU can be found before World War II in the U.S. and Europe, where prototypes were built. The *Shinano,* however, is the first fully developed train with tilting cars to commercially be operated in the world.

Its introduction during a time of serious labor union-management relations complicated negotiations on train schedule revisions even more. The most serious sticking point was not the actual operation of tilting EMU trains, but the fact that electrification of the Chuo Line put steam locomotives out of commission. This eliminated the need for steam locomotive inspections and repairs at Nakatsugawa Engine Depot. Most employees who had worked here were to be transferred to Jinryo EMU Depot in nearby Nagoya. Although they would not lose their jobs, workers who had been able to walk to work would now have an hour or so commute by train. This inconvenience prompted intense negotiations with workers opposing rationalization.

EMU malfunctioning during test runs immediately complicated labor union-management bargaining. The labor unions would push to postpone schedule revisions on the grounds that these trains were too dangerous to operate.

Although there were fewer mechanical failures with tilting EMU trains than there had been with newly developed EMU trains in the past, there were problems with the electrical system. The swaying of the tilting EMUs caused friction between the electrical cables linking the railcars, ruining the wiring insulation. This is the type of problem that can only be discovered over time with test runs.

The biggest shock came when one of the EMU trains nearly derailed.

I was at a bowling alley next to our company-owned apartment, trying to enjoy my time off on a Sunday afternoon close to the day when the schedule revisions were to be held. My pager suddenly went off, and I checked in with the command center to find that an EMU train had nearly derailed during a test run.[4] Scattered ballast had broken windows in nearby houses.

I rushed to the scene of the accident. The problem EMU train lay on the track in the same position it was in when it had finally come to a stop, with the bolster anchor from the bogie dislodged. It looked as if it had been used as a pole jump.

Fortunately, the train had not actually derailed, but the bolster anchor was bent limply, and the scattered ballast had been thrown into nearby houses.

An emergency labor union-management bargaining was held the following day.

The union insisted; "Our members would not operate a train as dangerous as the *Shinano*."

We told them; "Each EMU train had undergone spot checks, and steps had been taken to guarantee that bolts would not come loose." Still, they could not be convinced.

Management was finally able to defuse the crisis by angrily insisting, "Train schedule revisions would definitely be postponed if such a situation occurred again."

Experience in a similar situation as the Head of the EMU Division in Tokyo helped during this crisis. I was confident during the tilting EMU failure negotiations that there would be no additional problems. In Tokyo, when the explosion of an electric motor had caused failures on three occasions, I had been worried that I would be forced to resign if this actually did happen a fourth time.

Another problem with the tilting EMU trains was motion sickness. Since the EMUs tilt laterally as the train takes a curve, passengers with extremely sensitive stomachs experience something close to seasickness. This problem became apparent when some conductors complained of motion sickness during test runs. I rode on a number of test runs myself, but perhaps because I have a strong stomach, I never felt sick. I was shocked, though, to see a young woman who had ridden from Nagano on an in-car sales training test run slumped over and looking sick on the Nagoya Station platform.

Around this time, we staged a tilting EMU train test run for the press. Newspaper reporters were extremely interested in the problem of motion sickness. The reporter in charge of the local newspaper heard my explanation "The first tilting car train in the world brings us much closer to Nagano. The *Shinano* will make the run to Nagano in just three hours and 20 minutes. It will also eliminate pollution from steam locomotive smoke." This prompted a response from a local reporter that hit us exactly where it hurt, "We've heard that the terrible swaying of this train actually makes people sick."

I just pleaded my case.

The swaying is nothing to worry about. It actually feels nice, like dancing the Waltz.

---

[4] Before cell phones and smartphones became popular, pagers were in widespread use. However, these could only deliver simple data. Therefore, it was necessary to reply to a call from a fixed-line phone.

On the following day, when I went to work at the Railway Administration Bureau, everyone seemed to giggle when they passed me in the hallways. I figured out why when I saw the local newspaper in my office. The top story on the front page featured a photograph of the new EMU train under the glaring headline, "Waltzing EMU Train Runs." It contains a discourse article with my name on it. The press conference was successful in erasing the bad image of the tilting EMU train, but I couldn't believe that such a flippant comment would literally become the front page headline.

On July 10, 1973, ten days behind schedule, the *Shinano* limited express with its tilting EMUs was introduced. Although we had managed to convince the labor unions of the EMU's safety and were confident that there would be no problems, Headquarters had backed off at the last minute. Immediately before the train schedule was to be revised, its operation was postponed for ten days, as they justified it, sufficiently testing the train before its introduction. Although the tilting EMU train experienced few mechanical failures and was successfully introduced, the problem of motion sickness continued to haunt us.

Anyone walking through the cars of a fully booked *Shinano* at that time would notice the dirty lavatories in most of the cars. Passengers with sensitive stomachs apparently vomited in the toilets. Even though I have never experienced this myself, I am not overly fond of the way the EMUs tilt. The downward tilt of the car body is particularly uncomfortable, and it was around this time that I began to voice my opposition to these trains inside the company.

6.8 Ride Comfort not Measured with Data

Ride comfort is not only measured by the force experienced passing through curves. The swaying of the car body that makes passengers feel sick when rounding curves is what actually makes an EMU ride uncomfortable. Train car technicians, however, seem to believe that the magnitude of the lateral force is the only factor involved in passenger comfort when rounding curves. In reality, the Measurement Department is solely concerned with the magnitude of lateral centrifugal force when assessing the comfort level on test runs.

> The lateral force when taking curves is not the only factor for a comfortable ride. These assessments are a waste of time if they do not take the discomfort caused by the tilting EMUs into consideration when determining whether or not passengers will be comfortable. My objections have fallen on deaf ears.

I believe they are reasonable because comfort is not statistically measurable.

I once asked Masaru Ibuka (the deceased), the chairman of SONY, who had been asked to serve as Chairman of the Railway Technical Research Institute, how comfort should be measured.

He replied,

> It's the same as when testing the sharpness of a knife. Using statistical data is meaningless. A knife's sharpness can only be measured by the human senses.

His words prove that a man with a thorough knowledge of his profession speaks the truth.

To test this concept, we asked 100 people to take a ride on the Chuo Line near Kofu on an EMU train not equipped with tilting cars. The speed at which the train took the curves steadily increased, and the passengers were asked to grade their level of comfort during the ride. Their replies varied greatly. Some passengers complained about the swaying once the train reached a certain speed, while others enjoyed the feeling of speeding along the rails at exactly the same speed.

The situation is comparable to that in a car. Drivers who fly down the highway at top speeds are not likely to worry about the lateral vibration when rounding curves, and this may actually be their idea of comfort. The sensation of ride comfort varies from person to person but also depends on the particular conditions at a given time, the frequency of the lateral vibration, and many other factors.

Judgment must ultimately be left to the passengers themselves. Whether a greater number feel that faster is better, even with some vibrations, or dislike the vibration of the tilting EMUs. To test passenger reactions, certain Chuo Line *Azusa* limited express train services were equipped with tilting EMU trains in 1993. These trains have been relatively well received, but it is difficult to know whether their popularity is due to the faster speeds achieved with the tilting EMUs or the fact that the new design is attractive to passengers.

The *Azusa* incorporates slightly more advanced technology than the original *Shinano* EMU train. The idea with the *Shinano* was simply to use centrifugal force to tilt the car body on curves; the degree and direction of the tilt depended on the conditions at the time. This is called 'passive tilt'. Another method, 'active tilt' uses air pressure to tilt the car body before the train enters a curve. This type of car improves the comfort level of passengers somewhat by controlling the tilting degree of the car's body.

The active tilt method was first put into effect in Japan by JR Shikoku and is now used in *Shinano* trains, as well. The *Azusa* is also equipped with active tilt cars, which have changed the way the cars tilt, but made no difference to the fact that they vibrate. I would like to study passengers' reactions to these developments for a bit longer. Italy was the first country in Europe to adopt fully developed tilting cars. In 1988, the *Pendolino* (Italian. tilting car) EMU limited express was introduced between Milan and Rome. This section of track is comparable to Japan's Tokaido Line; it transports the largest number of passengers and serves as the most important line for the Italian State Railways (FS).[5] As such, it has stiff competition from the airlines.

The Montes Apenninus range, the backbone of the Italian peninsula, lies midway along this line. The tilting EMU limited express trains were adopted in order to deal with the significant number of curves along this mountainous section of track.

To compete with airplane travel, simple meals and newspapers are offered on the *Pendolino* free of charge. The tilting EMU limited expresses, however, have now been replaced with a new type of EMUs without a tilting system and the tilting EMUs have been moved to another section. This may result from problems with the tilting

---

[5] Now it's privatized.

motion similar to those Japan experienced. I have ridden this section twice, and the tilting of the cars in this mountainous area was quite uncomfortable.

Recently, however, the tilting car has become quite the rage in Europe. The improved Italian *Pendolino* operates across the Swiss and French borders, and the tilting car technology used in this train has been adopted by DB and VR. Sweden operates its own tilting EMU train, the *X2000*, which has become the Swedish signature express train.

The *Azusa* tilting trains operate at speeds of 70-100 km/h between Takao and Kofu due to a series of sharp curves that the train must negotiate.

In Sweden and Germany, however, trains take curves at a speed of 160 km/h. Geographically, these regions are considerably different from those of the Chuo Line, with curves that are not nearly as sharp followed by long straight sections. Passenger comfort does not present much of a problem under these conditions. Tilting cars take on a very different meaning in each country.

Shinkansen has also recently been equipped with tilting EMUs. This is not happening on the *Shinkansen* in Japan,[6] but is being used on high-speed trains in France and Germany. In 1998, France developed the tilting TGV *Pendular*, and Germany came up with the tilting ICET. Although these trains are only prototypes,[7] it is possible that they will be developed for full-fledged operation.

The track gauge for the newer high-speed trains, the equivalent of the Japanese *Shinkansen*, is the same as those for conventional trains in Europe, which enables the faster trains to operate with no adjustments on all lines. Tilting EMUs are needed, however, for these trains to operate through to particularly curvy sections of track.

There is, however, a quite political reason for these developments, as well. There was a time when the success of the TGV prompted France to formulate major plans for the construction of a national TGV network, similar to Japan's *Shinkansen* network. Despite the success of the TGV, however, the financial situation at SNCF was deteriorating, and the government could no longer afford to finance these plans in anticipation of a European monetary union. Local governments had been hoping that TGV network construction would bolster local economies; the new tilting EMUs were therefore used to appease local governments by quickly attracting attention to an alternative to the TGV plan.

Another rumor is that a concerted effort by Italian railcar manufacturers to sell the tilting EMUs they manufacture has been behind their recent popularity. These cars may have come to symbolize high-speed railway construction shaking up the political community.

In Japan, there have been some arguments for introducing a tilting system for mini Shinkansen EMU trains, but my personal opinion, as a railway man who has

---

[6] Since 2007, the Tokaido *Shinkansen* has introduced the N700 Series EMU, whose body is tilted by about one degree by air springs, to improve speeds when negotiating curves.

[7] DB has been operating the EMU ICE-T and DMU ICE-TD with hydraulic body tilting system (different from the natural pendulum system) since 1998.

experienced many unexpected problems, is that it would be better to avoid complicated structures for *Shinkansen* EMUs running at 270 km/h, even though extremely
timid.

Chapter 7
Race for Speed in Europe

In this chapter, the author introduces how the high-speed railways of France, Germany, Spain, Italy, Belgium, Russia, and other countries in Europe confronted the challenge of high-speed travel. He also explains the rivalry, local sentiments, and different stances toward technological challenges in each country, based on his actual experience working at UIC in Paris at the time, as well as on the opportunities he has taken to travel to Europe and exchange opinions with the people concerned. His unique critique from a different perspective from those involved in Europe is a must-read.

7.1 France Runs TGV at 270 km/h

The first section of the high-speed TGV railway in France connecting Paris with Lyon opened on September 27, 1981. The new line, however, had not then been completed, and TGV service began on only half of this line. This train reached a maximum speed of 260, 50 km/h faster than the Japanese *Shinkansen* at that time.

When the entire 425 km line was completed two years later, a maximum speed of 270 km/h was achieved.

We were naturally aware that the TGV was under construction and knew that France's goal was to operate a faster train than Japan had been able to build, so the news came as no surprise. Rather, the sense in Japan was that the inevitable had finally happened—the front-runner had taken her place in the field. Since the end of World War II, France had led the world in railway speed technology.

SNCF seems to me a bit too passionate on the topic of high-speed operations and overly proud of its accomplishment. Still, speed is definitely the primary and most obvious indicator of technical achievement and is one key to increasing railway competitiveness.

© Japan Railway Technical Service 2024
S. Yamanouchi, *If there were no Shinkansen*,
https://doi.org/10.1007/978-981-99-8890-7_7

While working on the development of the *Shinkansen*, Japanese technicians studied and incorporated French railway technology. Younger technicians went to France as exchange students to study its railways. A latecomer to the railway community, Japan went on to develop the *Shinkansen* and set a 200 km/h world record for railway speed. This must have come as quite a shock to the French railway technicians.

A friend at SNCF, whom I came to know around the time I began working at the Paris UIC Headquarters in 1969, summed up the reaction like this:

> There is no technology used in the Japanese *Shinkansen* that one can point to as innovative, is there? If your accomplishment is in constructing new lines, that can be done here at any time we choose. The French had no interest whatsoever in the Japanese railway system. It was as if it didn't even exist among the world's railways. That is until the *Shinkansen* was developed. Now, France catches a cold whenever Japan sneezes.

This conversation took place around the time that the French were becoming interested in Japan, in its economy and industry in particular. One French television program at the time featured Japan's modern car and camera factory lines and its *Shinkansen*. The end of the program showed a middle-aged worker at a clock factory in central France that had closed down; with a lonely expression, he was lamenting, "France had lost everything to Japan."

Soon after we arrived in Paris for my stint at the UIC, I took my family to Switzerland by train. An elderly couple sharing our compartment were carrying a copy of the newly published "Japan, The Third Power" written by Robert Guillain. They threw open the book, spoke profusely about their admiration for my country, and fired off question after question about Japan.

SNCF was technically capable of operating high-speed trains at 200 km/h. Unfortunately, they possessed no lines that could accommodate a high-speed train like the *Shinkansen*. To compensate, they chose the conventional lines with the fewest curves to be subjected to the high-speed conditions of the 200 km/h test.

In May 1967, the first 200 km/h operations in Europe began with the introduction of *Le Capital*, a limited express train operating between Paris and Toulouse. The only time the train actually reached this speed, however, was during the 80 km section between Orléans and Vierzon.

I had the opportunity to accompany Hideo Shima, the former JNR Chief Engineer, on his ride in the driver's cab of the locomotive hauling this train. After the train had passed Orléans about an hour out of Paris, it quickly picked up speed, climbing to its maximum 200 km/h speed. The driver of the locomotive seemed a bit nervous, perhaps because the man who had fathered the *Shinkansen* was standing next to him. The speedometer needle stayed at 200 for quite a while. Being in an 82-ton locomotive traveling at this speed is a powerful feeling, much more intense than in the relatively lightweight *Shinkansen*.

Unlike Japan's high-speed lines, this train traversed level crossings and turnouts and passed small provincial stations at 200 km/h. A shrill alarm is activated periodically to prevent drivers from dozing off. It was a very tense, high-speed driving experience (Fig. 7.1).

Fig. 7.1 French TGV trains at Charenton Depot in Paris. *Photo* provided by Katsuji Iwasa

The next country to achieve speeds of 200 km/h was the United Kingdom, the birthplace of the railway system. UK had been passionate about operating high-speed trains before World War II and had even set steam locomotive speed records. After the war, however, the British government was under financial pressure and reluctant to extend large-scale capital investment to its railways, and the country was slow to electrify its lines, much less develop a high-speed railway system. By this time, however, the UK had built its high-speed train (HST), a diesel train that reached speeds of 200 km/h. The HST was introduced in 1976 between London and Bristol and then on other lines.

One year before France introduced its TGV trains, Germany had introduced 200 km/h operations between Munich and Augsburg. Rail wear proved worse than expected, however, and operations were suspended soon after the federal government recommended that they be stopped. Full-fledged operations at this speed began here in 1978.

The record speed of 200 km/h was a great achievement, but trains could operate at this speed on only a few sections of line. This was not a success in the true sense of the word, but rather a contest of pride. A completely new railway designed specifically for high-speed trains was needed to fully develop a high-speed railway system.

7.2   *Shinkansen* **Ignites High-Speed Dreams**

The success of the Tokaido *Shinkansen* in Japan kindled the world railway community's dream of achieving high-speed railway systems. 'The TGV Challenge' was published in France in 1981. In the chapter entitled 'Son of Tokaido,' the author describes the impact,

> The true mother of the high-speed railway may in fact be France, but the Paris-Lyon TGV line is the child of the Tokaido *Shinkansen* trains and the cousin of Italy's *Direttissima*. SNCF was very interested in the developments in Japan and kept a close eye on the research being done in the 1950s and on the developments since its opening and operation during the 1960s. Japan has greatly impressed both its collaborators and rivals by adopting a completely different system that defied common sense. For example, lines were designed specifically for limited express trains and perfectly straight lines were laid for high-speed operations. This concept instantly sparked interest and captured people's imagination.
>
> If SNCF is at all inclined to remove its kid gloves, it must immediately step up to bat.

I think I fully understand the impact that Tokaido *Shinkansen* had on SNCF officials.

This book also describes other aspects of *Shinkansen* operations.

> One is truly struck by how they are managed. Operations are highly concentrated, blending *Hikari* and *Kodama* trains and incorporating the first lines in the world with no signals. Trains are equipped with a series of continuous cab signals, CTC devices, ATC devices, and so forth and on. While not exactly crude, passenger service is quite plain. The cars are designed with rows of two seats on one side of the aisle and rows of three on the other. Vendors pass through the cars selling food and beverage ceaselessly. The Tokaido *Shinkansen* is mass transit in the true sense of the word.

I was taught the hard way about how the French railway community feels toward the Japanese *Shinkansen*.

In the autumn of 1971, if I recall correctly, I was working at the UIC Headquarters in Paris when a conference was held by the French Railway Industry Council. The guest speaker was Pierre Sudreau, Council Chairman and, as former Minister of Education, a dominant figure in politics. He also served as Chairman of the France-Japan Friendship Society. His speech that day, however, was anything but friendly.

> Half of the Christmas cards I receive from Japan are adorned with pictures of the *Shinkansen*. Japanese people are using these cards to promote the *Shinkansen* around the world. This cannot be forgiven. France must build her own *Shinkansen*.

This was met with enthusiastic applause.

Toward the end of the year after the Tokaido *Shinkansen* had been introduced, the North Branch of SNCF conceived its own high-speed railway concept. They gave this scheme the name 'TGV' (ultra-high speed railway). This marked the beginning of the TGV system.

SNCF established a new organization, the Research Bureau, the following year, and de Fontgalland who was later appointed UIC Secretary-General, was selected as its first General Manager.

The TGV development plan was officially introduced with the 'C03 Project' on July 10, 1969.

This project differed from the Japanese *Shinkansen* in that it was based on the concept of operating high-speed trains directly through to the conventional lines. This structure would offer a more convenient system for transferring passengers and eliminate the need for extremely costly construction in the heart of urban areas.

Construction expenses would be further reduced by making the gradients considerably steeper than those on *Shinkansen* lines. Despite the short distance between Paris and Lyon, the TGV gradient would be a relatively steep 35/1,000. The steepest Tokaido *Shinkansen* gradient near Sekigahara is only 20/1,000.

In June 1968, Vienna in Austria hosted an international symposium on speeding up of railway systems. It was here that Research Bureau General Manager, de Fontgalland outlined the concept for operating TGV trains on the approximately 500 km between Paris and Lyon, the second largest city in France.

Unlike the significant number of large cities in Japan, Lyon and Marseille are the only two French cities with a population of one million or more, with the exception of the Paris metropolitan area with a population of 10 million. 40% of the country's entire population lives within an area encompassing Paris, Lyon and Marseille and, to the East, Cannes and Nice in the Côte d'Azur region along the Mediterranean Sea. This railway line running through this area is comparable to the Tokaido Line in Japan. This made it a natural choice for the development of the TGV system.

The plan called for a line connecting Paris with Lyon in as straight a manner and short a distance as possible with no intermediate stations. This would reduce the previous 512 km route to 425 km in one fell swoop.

A movement grew out of opposition to the plan in the historical city of Dijon, which lies between Paris and Lyon. TGV detractors in Dijon found it unforgivable that the TGV line development plan neglected their city. Dijon is only a mid-size city with a population of 230,000, making it the 25th largest city in France, but it is the heart of the region that produces the world-famous Bourgogne wine. It was also the capital of the medieval Duché de Bourgogne and was the site of the battle between the British royal family and the French royal family led by Jeanne d'Arc.

In opposition to the TGV plan, the Dijon Chamber of Commerce submitted a petition advocating highway and canal development to the national government. Frederique Rodet, an elderly technician living in Dijon, had his own ideas. He called for the development of an air levitating train 'Aero train' between Paris and Dijon.

SNCF firmly opposed these alterations to its line development plan, because a route passing through Dijon would take TGV trains an additional twenty minutes.

Robert Poujade, then Minister of the Environment, was elected as Mayor of Dijon in March 1971. Simultaneously serving as a mayor and a congress member concurrently is not prohibited in France, but it put Poujade in a difficult position. As the city's mayor, he would be required to represent the interests of Dijon residents. A considerable portion of his electorate, however, was made up of railway workers, and he would not want to be an enemy of SNCF. Mayor Poujade responded to this situation at a City Council meeting on November 6, 1972.

This is an extremely complex problem … I do not think it advisable to draw any rash conclusions. Developing the TGV system is not so easily done. I believe it will take ten years or more.

The TGV plan faced difficulties in Paris, as well. Although the government had begun studying this plan around 1970, Prime Minister Chaban-Delmas was never whole-heartedly in favor of it. As he was apt to do, Minister of Finance Giscard d'Estaing, later elected President of France, put forward a number of problems concerning the TGV.

This Minister was nicknamed "Yes, but … Giscard" by the French people. He was a stereotypical intellectual official who, while never opposing anything outright, proceeds to find any number of problems with whatever is put before him.

"Now, how much will construction cost? Is this not simply a hobby that the technicians enjoy working on like the Concorde was? Wasn't it developing the *Shinkansen* that threw JNR into debt?" etcetera, etcetera…

Desperate efforts on the part of SNCF finally pushed the Sixth Long-Term Plan through in 1972. It suggested,

Based on projections that railway operations between Paris and the Southeast would have reached saturation point by the end of the 1970s, so they would begin preparing to construct new lines between Paris and Lyon.

No unusual technical problems plagued TGV development, but financial efficiency was a problem. SNCF submitted data indicating that investment returns would be 17% and government subsidies would be unnecessary. The success of the *Shinkansen* in Japan also provided extremely convincing material for the railway.

André Sègalat, then Chairman of SNCF, visited his friend President Pompidou at the Elysee Palace frequently to advocate for the TGV plan.

A compromise was even proposed by the Dijon government. It is to construct a route that will branch off from the middle of the TGV main line to allow access to the conventional line to Dijon with two intermediate stations built in the Bourgogne region. This Dijon branch line was nicknamed the 'Poujade branch line' by the public.

The plan for an Aerotrain also disappeared into obscurity, remaining only a dream when its creator Jean Bertin died.

7.3   Gas Turbine Cars Face Extinction

Paving the way for TGV construction was Cabinet approval on March 6, 1974. TGV development was the final resolution enacted before Pompidou passed away in the middle of his term as President. Construction began along the Paris-Lyon section on December 7, 1976.

In 1967, ALSTOM, the leading rolling stock manufacturer in France, had brought the concept of using gas turbine railcars on high-speed TGV trains to SNCF.

SNCF had operated gas turbine traction units on conventional limited express trains since the previous year. Gas turbine cars are more powerful than their diesel cars and are effective in increasing speed on non-electrified lines. The ALSTOM concept was to use this technology to develop high-speed railcars for TGV trains.

The French do not like imitating other countries, preferring to pursue their own unique concepts and technology. The gas turbine TGV car illustrates this philosophy. It eliminates the burdensome problem of developing technology to collect electricity from the pantograph at high speeds. They had faced a problem during their 1955 high-speed test run when a pantograph had burned out. Gas turbine cars also eliminate the need for electrification, which brings construction costs down. This improves the economic feasibility of the system, which is the government's greatest concern. These cars can also be operated on sections of lines that have not been electrified. It was for these reasons that SNCF had decided to use the gas turbine method to power its TGV trains.

I answered the telephone in my UIC Headquarters office; I believe this was in the autumn of 1971. The voice on the other end announced, "My name is Fernand Nouvion and I would like to meet with you."

Fernand Nouvion had led the development of the AC electrification process and had overseen the 1955 trial runs. Even in Japan, a considerable number of technicians worshiped this man. He had been a 'Worldwide Electric Railway God.' I gladly accepted his invitation and joined him for lunch at the Hilton Hotel next to my office.

By then, he had already retired from SNCF and moved to a rolling stock manufacturer but was extremely youthful and suave. He was a man of no distractions.

> The Japanese *Shinkansen* is a wonderful train. I myself have ridden it numerous times on my visits to Japan, and it is truly a technician's dream come true. We have the *Shinkansen* to thank for revitalizing passenger transport on railways around the world. SNCF is saying that it will use gas turbine cars on its TGV trains. What do you think about this? I personally am opposed to this idea. Gas turbine cars would fill the stations in Paris and Lyon with exhaust and its thirmal efficiency is not high.

As an engineer who promoted AC electrification, he spoke passionately about the railway. About the time that we had finished our meal, he concluded,

> May I speak frankly? There is one thing that I do not like about the *Shinkansen*. Too many pantographs on those EMU trains.

The *Shinkansen* comprise 16 cars running on eight pantographs. The high-speed trains in France use DC electric locomotives with considerably more current being drawn than on the *Shinkansen* and only require one pantograph.

The gas turbine TGV001, the first TGV test car, was completed in March 1972. The lavish exhibition held at Montparnasse Station in Paris on June 8 was even attended by Minister of Transport Jean Charmant.

I also visited this exhibition, finishing up my three-year stint at UIC the following day and returning home from France.

The 1973 oil crisis, however, hit the TGV development plan hard. It was also the fatal blow to gas turbine cars with their poor fuel consumption efficiency. SNCF hastily shifted the basis of its TGV development plan to electrically powered trains.

Nouvion's wish had been granted. Following the Z7001 prototype, two prototype test trainsets for practical use were developed and nicknamed *Patrick* (TGV01) and *Sophie* (TGV02). Then began a series of high-speed test runs.

On February 26, 1981, more high-speed test runs were conducted with railcars for commercial use before partial TGV service was introduced. During these runs, TGV trains hit a new world record of 380 km/h, the first time a railway speed record had been broken in the 26-year period since 1955.

On October 1, 1983, construction of the entire TGV line between Paris and Lyon was completed. This was two years after the partial service had been introduced in 1981. President Mitterrand's attendance made for a spectacular opening ceremony. Maximum TGV speed was also increased from 260 to 270 km/h, and the three hours and 50 minutes it had previously taken to get to Lyon from Paris was cut to a mere two hours.

7.4   *Shinkansen* and TGV: The Battle for Superiority

This time, the news of France's TGV achievement of 270 km/h shocked Japan.

People worried that "Had the *Shinkansen*'s position as the world's best train been snatched away?" or "Japanese technology was now inferior to the level in France."

The actual situation, though, is not so clear-cut. Nineteen years had already passed by then since the Tokaido *Shinkansen* began service. It is obvious that newer ones are more advanced. We also have to ask whether the technical level of a system should be judged by speed alone. Speed is definitely a powerful indicator of technical skill, but technology involves many more elements than can be measured by a single index.

With the successful development of the *Shinkansen* in 1964, many people happily claimed that Japan now possessed the world's finest railway technology. Even then, I doubted whether we could really claim to be the best.

There is no doubt that the *Shinkansen* was an incredible train or that it was the fastest in the world in its time. Its railcars, tracks, and signals all incorporated the latest technology. Most of this technology, however, was originally developed in Europe. Japan improved on these techniques as best suited its needs, but did not develop most of them on its own.

What Japan did do, however, was to visualize the concept of the *Shinkansen* and use this concept to formulate a system within a short amount of time. The technology that made this into reality is indeed wonderful, and it is one aspect of the technologies. French technology values originality, while Japanese technology is particularly skilled at putting techniques to practical use and improving upon them. These characteristics are reflected in the TGV and *Shinkansen*, respectively. Another valuable aspect of technology is the development of a thorough knowledge of safe, precise operations and maintenance. JNR accomplished this by working through the numerous problems it faced after the developed *Shinkansen* was in operation. Taking all of this into consideration, who can say that either Japan or France has achieved a greater level of technical skill?

. When I saw the completed TGV technology, the technical philosophy behind it was somewhat shcking to me. They seek theoretically sound technology. They never imitate the Japanese *Shinkansen*. This is the manifestation of a burning desire to build a system best suited for France.

Let's take a look at some specifics. Every *Shinkansen* railcar is EMU, a distributed traction system that has electric traction power in all cars. In contrast, the French TGV employs power cars that carry no passengers, coupled to the train on each end. This is called a concentrated traction system.

This difference is the result of a pragmatic stance by JNR toward operating rapid trains on weaker infrastructure. We were not concerned with theoretical soundness. Moreover, railways in Japan transport an exceedingly large number of passengers, and this requires trains to accommodate as many passengers as possible. Converting the *Shinkansen* to a train with power cars would reduce capacity by two cars' worth; this would create serious difficulties.

In Europe, however, conditions are entirely different. The ground is solid; the tracks are stable; the railway does not transport such large numbers of passengers. Locomotive advocates assert that "trains with power cars are less expensive, because of the need for fewer power units." This argument, however, is entirely theoretical; no one knows for certain whether this would actually prove true.

Converting the Japanese railway system to trains with power cars at this point would be extremely costly, I believe. Not only because we do not have a thorough knowledge of high-speed power car technology, but because manufacturing facilities and maintenance depots have been designed for EMU trains. The advantages of trains with power cars do not seem great enough to offset all of the problems conversion would entail.

European countries would also likely face the same type of problem in converting to EMU systems. DB did, in fact, build the ET403, an EMU limited express train, but it was not particularly successful. When I asked a DB executive why this train had been abandoned so quickly, they replied "the cost and maintenance expenses were prohibitive."

7.5 Solid and Deliberate France

France did not imitate Japanese *Shinkansen* technology in developing its TGV. SNCF persisted with its traditional trains with power cars. A single locomotive is not powerful enough, and it is difficult to move a locomotive to the other end of a train during shunting. The railway dealt with this by coupling a locomotive at both ends of a train. Despite these adjustments, the trains still lacked power; to increase their power, an electric motor was added to the passenger car's bogie behind the locomotive. It seems to me that they don't have to stick to the locomotive hauling method, but French people place a high value on theory.

The TGV and *Shinkansen* differ slightly in terms of passenger car design. Each car in a *Shinkansen* EMU train is supported by two bogies. The TGV has only one bogie mounted between two passenger cars. This arrangement is termed 'articulated,' and the Odakyu limited express, nicknamed the 'Romance Car,' has used this design for many years. In theory, articulated cars experience less swaying and are more stable when running.

JR East conducted a comparison test by coupling cars of both structures to a test EMU train called *Star 21*, but there was no noticeable difference. France's decision to adopt articulated cars illustrates a difference in philosophy. Japanese tend to stick with "If there is not much difference, then the system shall remain the same." The French TGV philosophy, on the other hand, was "To adopt the theoretically better method."

The TGV also employs a unique type of car suspension. On *Shinkansen*, car bodies are placed on top of conventional bogies, while on TGV trains, the bogie supports the car body at a fairly high position in the gangway between passenger cars. Theoretically, support from above improves stability. The use of articulated cars on the TGV makes this design possible.

The *Shinkansen* employs air springs, but metal coil springs were initially used instead on the TGV. Before TGV development began, SNCF imported Japanese air springs for testing, but they were not adopted at this time.

While the difference between metal coils and air springs is not so large in terms of passenger comfort, air springs were not adopted because coil springs do not consume large volumes of pressurized air, eliminating the need for large air tanks. This decision was, I imagine, also affected by the rivalry the French felt toward Japan, which had already adopted air springs.

Significant vibration on TGV trains later forced SNCF to replace the coil springs on all cars with air springs. Air springs are said to be imported from Japan.[1]

For a train with power cars, the TGV uses few pantographs. Only one pantograph is equipped on the power car at each end of the train. It is normally running with one side down. There is only one pantograph per train. Japan has recently reduced the number of pantographs it uses on the *Shinkansen*, but these trains still require a minimum of two.

In this regard, Japan feels that "There are a lot of the various troubles, so it runs with two for now. Not that big of a difference." The French philosophy, on the other hand, is "the fewer the number of pantographs used, the better."

The arrangement of the overhead contact wires transmitting electricity to the pantographs is also different. The *Shinkansen* was originally developed with a complex arrangement called composite compound overhead contact wires. This

---

[1] Sumitomo Metals of Japan put air springs to practical use and commercialized them as 'Sumiride.' The dining car of the TEE train *Mistral*, which appeared in 1969, used the 'Sumiride' type in order to adjust the height by means of air springs. The 'Sumiride' air springs were installed on the TGV001 gas turbine test train set, which was completed in 1972. However, air springs were not initially used in the trainsets used for commercial operation of the TGV, which was introduced in 1981. Later, however, air springs were adopted from the standpoint of ride comfort. However, these were made in Europe. It is thought that this is because the patent term of 'Sumiride' had expired and the air spring manufacturing capacity had been established in Europe. References; Seiichi Nishimura and Machi Nakata, No. 92 Rolling Stock & Technology.

complex arrangement reduced sparks to a minimum when EMU trains were running at high speeds.

The TGV overhead contact wires, however, are designed in a simple arrangement that does not differ greatly from that used on conventional lines. Japan has adopted a simpler arrangement similar to the French one on its new *Shinkansen* lines.

There are also differences between the two systems in terms of track structure. Japan began to shy away from laying ballast on top of earth roadbeds after we experienced numerous problems with the Tokaido *Shinkansen* and its time-consuming and costly maintenance. The tracks for the Sanyo *Shinkansen* lines were therefore laid on top of concrete structures.

The French used the traditional method involving laying ballast on top of earth for the TGV tracks. The obvious advantage with this type of track is its lower construction cost.

There is also a basic difference between Japan and France in terms of how the ATC safety system is used. On the *Shinkansen*, the ATC device automatically activates a train's brakes when a lower speed warning is displayed. We prefer machines to be in control of our safety. We think it's safer.

The principle in France, however, chooses human judgment over machine control. The idea is that the train is entirely operated by the driver, and the ATC device activates at times of human error. The driver is given as much responsibility and freedom as possible, and safety devices are used only when trains are operated in a dangerous manner.

Who is to say which of these two methods is better?

Train schedules differ, as well.

Since its introduction, the *Shinkansen* adopted a 'patterned train schedule.' According to an extremely straightforward and easy-to-understand timetable, *Hikari* and *Kodama Shinkansen* trains depart at specified times. This type of schedule was not, however, adopted by France for its TGV trains. Trains run more frequently during times of peak demand and less often when there are fewer passengers. The idea behind this is that passengers will consult the timetable to choose the train they will board, and there is some truth in this.

Few passengers memorize even the simplest train schedule; I consult the timetable nearly every time I use the *Shinkansen.*

The *Shinkansen* timetable has recently become slightly more complicated so that the term 'patterned schedule' may no longer be applicable to the Japanese system as the railways try meticulously to meet the needs of all of its various passengers.

While, France has recently begun publicizing its 'equal interval TGV timetable.' The concept transition on both sides is interesting.

Another fascinating aspect of TGV development is how solid, in fact almost deliberate, France was during the realization process.

SNCF developed its first gas turbine prototype, the TGV001, in 1972, nine years before service was introduced. This was perhaps a detour since they subsequently shifted the premise of TGV development from gas turbine to electric motive power after the oil crisis. First, the prototype Z7001, a remodeled version of EMUs for conventional lines, was built, and a series of high-speed power collection tests were conducted.

Next, three years before the opening, the prototypes *Patrick* and *Sophie* were built and continued to undergo final testing. Then, when the trainsets for commercial operation began to appear at the end of 1980, they were thoroughly run before the opening of the new service.

Test run records show that the prototypes were tested for approximately 2.4 million kilometers, while the trainsets for commercial operation were tested over 4.6 million kilometers. The total number of test kilometers run was seven million kilometers. The TGV does not involve any entirely new technology. A system using wheels and rails that had been around for 150 years was simply improved and the running speed increased. Nonetheless, I'm impressed that they've done so many trial runs.

By way of reference, two different prototype EMU trainsets were test run on the Odawara test section for 250,000 km, while approximately 750,000 km in training runs with commercial EMU trainsets were conducted before the *Shinkansen* service was introduced. This totals one million kilometers for *Shinkansen* testing.

While it is inspiring, in one sense, that a railway that had developed only meter gauge tracks would be the first in the world to operate a train at 200 km/h with so few test runs, it is also natural under these conditions that so many problems cropped up after service was introduced.

The French TGV system is the product of efforts to overcome certain difficulties with theoretically sound techniques and to compensate with conscientious testing. The Japanese *Shinkansen* system was developed quickly by adopting only proven technology and avoiding potential risks. This is another difference, I believe, stemming from the different approaches to technology in Japan and France.

7.6 Italy: First European High-Speed Railway

The French TGV was not the first high-speed train to begin carrying passengers in Europe. This accomplishment, in fact, goes to Italy.

Comparable to the Tokaido Line in Japan and the Paris-Lyon Line in France, the main Rome-Milan Line is FS's most important section of track. It services an extremely large number of passengers and passes through the large Italian cities of Florence and Bologna. There was no longer much room for increasing the number of trains operating on this line. As with the lines in Japan, the Rome-Milan Line is laid through many mountainous areas with a considerable number of curves, as well as sections unable to accommodate speeds of more than 90 km/h.

Construction of the high-speed *Direttissima* line between Rome and Florence began in 1970, six years earlier than the TGV in France. In Italian, *Direttissima* is the superlative of the Italian adjective meaning 'straightest'.

Train service was introduced on the southern half of this section between the suburbs of Rome and Citta di Pieve, 122 km north of Rome, on February 24, 1977. Although only a partial introduction of service, it was the first high-speed railway line to be operated in Europe, being established four years earlier than the introduction of the partial TGV service. The entire section between Rome and Florence was completed in 1990. It had taken 20 years to complete.

The delay had to do with political problems. Arezzo, a mid-size city with a population of 90,000, lies near Florence and is known for its famous fresco by the fifteenth century artist Piero della Francesca. Initial plans did not call for the new *Direttissima* line to go through this city, However, the people of Arezzo opposed the plan that *Direttissima* did not go through this city. Amintore Fanfani, the most powerful political figure in Italy at that time, was born in this city. A proposal was finally enacted under which the high-speed line would detour considerably to the east near Florence to pass through Arezzo. The *Direttissima* line had literally been thrown a curve by the politicians.

Florence and other large cities are considered the cultural heritage of Italy, making it nearly impossible to build new railway lines through them. Planners decided therefore to build an underground high-speed line in Florence. FS hammered out the 'Interconeshioni,' the concept under which the *Direttissima* line would be constructed.

This term means 'interchangeable.' Ten junctions would be used as interchanges along the high-speed line between Rome and Florence. Trains stopping at intermediate cities would exit the high-speed line at these points, stopping at stations on conventional lines and returning to the high-speed line in areas outside of these cities.

A variety of trains, not only the sleekly designed Eurostar ETR500 limited express, operate on the *Direttissima* line in Italy. As a pun on the Italian dish 'Antipasto Misti,' a mix of various hors d'oeuvres, this line could be called 'Treni Misti,' meaning a mix of trains (Fig. 7.2).

Fig. 7.2 Trenitalia (successor of FS) 'Frecceiarossa1000/ETR1000' high-speed EMU trainset (entered service in 2015) at Rome Termini Station in August 2019. *Photo* provided by Anthony Robins

7.7   German ICE Breaks 400 km/h Speed Barrier

Germany, of course, could not sit by and do nothing while France developed its TGV train. This country, however, has no concentrated economic and cultural center like Tokyo or Paris. Berlin is the largest city, but the country was still divided into East and West, and Berlin was in the East at the time. These conditions made selecting a location for a new high-speed line difficult. In the end, it was decided that one line would run north–south along the border between East and West Germany between Hannover and Würzburg and another would link Mannheim and Stuttgart in southern Germany. Construction was begun in 1976, the same year as the French TGV, and was partially completed in 1979.

Unlike the low population in France, Germany is similar to Japan in that it is quite densely populated. Another similarity was the intense opposition to the environmental problems that made the construction of a high-speed railway difficult. The two planned sections were completed in 1991. The ICE (Intercity Express), a new type of limited express train that was developed specifically for the new high-speed lines, also began carrying passengers in that year.

In the same manner as the TGV, the ICE is a high-speed train with power cars that runs at a maximum speed of 280 km/h. Car interiors, however, are much grander than the TGV, and ICE trains also have full-scale dining cars. These extras, I imagine, are the result of DB's rivalry toward the TGV.

An ICE prototype set a speed record of 407 km/h during a test run on May 1, 1988. With this, Germany not only broke the record held by the French TGV, but also broke through the 400 km/h railway speed barrier for the first time ever.

The ICE runs on a network linking such major German cities as Hamburg, Berlin, Frankfurt, and Munich. Few sections on the high-speed line are yet complete, making it necessary for these trains to run mostly on conventional lines. On these lines, the train reaches a maximum speed of between 160 and 200 km/h. Germany plans to complete a new line between Cologne and Frankfurt along the River Rhine that is now under construction in 2001. This line is expected to reach 300 km/h and will employ an entirely new type of wireless signal system.[2]

The European high-speed railways run in close connection with airplane flights. This is primarily to encourage passengers arriving on international flights to transfer to the railways. France's TGV line has already been extended to Charles De Gaulle Airport in Paris and Satras Airport in Lyon, and magnificent new stations have been built at these airports. A large airport station is now under construction at Frankfurt Airport in Germany, as well.[3]

In addition, Germany is considering extending its high-speed railway operations beyond the ICE to include freight and other trains. This concept is very similar to the idea of Italy's *Direttissima* system (Fig. 7.3).

---

[2] This section opened in August 2002. The signaling system is the German proprietary LZB (Linienzugbeeinflussung) system, which allows for operational control similar to moving blockages.

[3] Frankfurt Airport Station for Long-Distant Trains (Frankfurt Flughafen Fernbf) opened on 30 May 1999.

Fig. 7.3 ICE1 on the right and ICE3 with the coupler cover open on the left at Frankfurt Central Station. *Photo* provided by Kuniaki Mori

7.8 Spanish High-Speed Lines Use French Design

France and Germany are not the only countries in Europe to operate high-speed railways. Spain developed a high-speed railway line between Madrid and Seville that was completed in time for the Seville Expo and opened for service in 1988. Although slightly different in design, the trainsets used on the Spanish railway are nearly exact replicas of those used on France's TGV.

ALSTOM, the company that manufactures TGV trainsets, produces these trainsets as well. In contrast to the emphasis on function in the interiors of TGV and *Shinkansen* trainsets, Spanish high-speed trainsets, while not quite as luxurious as Germany's ICE, are designed to imbue passengers with a sense of leisurely travel.

After its struggles with the restrictive meter gauge of its railway lines, Japan built its *Shinkansen* on standard gauge tracks. Spain, however, faced the opposite problem; its track gauge was excessively broad. This broad gauge complicated non-stop operations over the country's border with France.

To deal with this, Spain devised a technique that would automatically adjust the gauge width of a railcar's wheels during train operations. This has enabled Spain to run a small number of limited express and all sleeper overnight trains to Paris and Montpellier, the university town in southern France, without requiring passengers to change trains at the border since the end of the 1960s.

Although there has been talk in Japan, too, about adopting this method, the technology is currently applicable only on TALGO passenger cars with particular arrangements. Technology that would allow for the free adjustment of wheels with tractive power of a rolling stock, such as locomotives and EMUs, is still in the experimental stage.[4] The most fundamental aspects of safety are involved, and we must proceed cautiously when applying it to the *Shinkansen* and other high-speed trains. This technology should be used first on low-speed trains and only applied to high-speed trains once we have gained enough experience with it.

Spain's broad gauge tracks measure 1,668mm, 233 mm wider than standard gauge lines. The commonly accepted theory is that Spain, having suffered under the invasion by Napoleon, specifically built tracks of a different width than France to frustrate other invasions by the French army. No one knows how accurate this story really is. At any rate, Spain chose standard gauge tracks for its new high-speed line.

This makes the free interchange between high-speed and conventional lines, as seen in Germany and Italy, impossible in Spain. In the same manner as the Japanese *Shinkansen* system, the two sets of lines are completely independent. Spain chose a narrower gauge than they had used in the past to make non-stop operations onto France's TGV lines possible in the future. Since completing its Madrid-Seville section of the line, Spain has developed plans to construct high-speed railway lines between Madrid and Barcelona and ultimately cross the border to link up to France's TGV lines.

The fact that the variable gauge bogie technology of TALGO is used on Spain's high-speed railway enables trains to pass directly from standard gauge high-speed lines onto broad gauge conventional lines is extremely interesting. This eliminates

Fig. 7.4   Talgo 350 *AVE* train (right) and Regional Train (left). *Photo* provided by Keiji Musha

---

[4] Subsequently, research and trial production of variable gauge EMUs was conducted by a research association, but various problems have yet to be solved.

the need for passengers to change trains in Granada, Málaga, or other cities and is, perhaps, one of the Spanish high-speed railway's defining features.

In terms of speed, however, the variable gauge TALGO trains are capable of reaching a maximum speed of only 200 km/h, as opposed to the 300 km/h at which the *AVE* (Alta Velocidad Española) trains, which are designed specifically for high-speed lines, can run[5] (Fig. 7.4).

7.9   Belgian High-Speed Railway Plan Encompasses the Whole of Europe

In 1998, service on a new high-speed line was also introduced in Belgium. Branching off from the high-speed line running between Paris to London via the Channel Tunnel, this line carries passengers to Brussels, the capital of Belgium and the location of the EU Headquarters. With the introduction of service on this new line, passengers are shuttled between Paris and Brussels in just one hour and 25 minutes on an equal interval schedule like the Japanese *Shinkansen*. The country plans to increase the maximum speed to 320 km/h in the near future.[6]

Additional plans have been developed to further extend this high-speed line as far as Cologne in Germany and Amsterdam in Holland.[7] European integration has shifted the focus of high-speed railway plans away from a single country, and more and more of these plans encompass the entire European region.

The TGV train operating between Paris with Brussels has been dubbed the *THALYS*. There is no dictionary listing for this word. When asked what it means, an SNCF executive explained,

"It means nothing in particular. We could not reach a consensus on any name related to a word used in France, Germany, or Belgium, so this name was chosen."

The German–French rivalry is extremely intense. SNCF rejected the idea of Germany operating its ICE train in France on the grounds that it was not up to its railcar standards. This issue was debated for two years before an agreement was ultimately reached.

The high-speed railway now being built by DB between Cologne and Frankfurt has a gradient as steep as 40/1,000 in some places. SNCF took offense, saying that DB was trying to prevent the TGV from operating on this line.

In order to develop trains that can run on this steep gradient, DB abandoned traditional trainsets with power cars and has been working on the development of an EMU train, the ICE3. This EMU is also designed to run on French TGV lines so that each axle load is reduced to only 17 tons and can run on the French electric power system. In Europe, the electrical systems of locomotives and EMU trains differ

---

[5] The Talgo 250, introduced in 2006, has a maximum speed of 250 km/h on standard gauge sections and 220 km/h on broad gauge sections.

[6] As of July 2023, the maximum speed remains 300 km/h.

[7] This plan has not been realized.

according to country. France employs a direct current of 1,500 V and an alternating current of 50 Hz/25 kV; the Belgium system involves a direct current of 3,000 V; while Germany uses an alternating current of 16 2/3 Hz/15 kV.

The differences do not stop with the electric power system. The signal systems are also diverse. All of these issues complicate international train operations. In Europe, this problem is referred to as 'inter-operability' and is one of the most serious issues that European railways face in the future.

Another example of European rivalry involves France and Poland. In June 1995, a grand ceremony commemorating the 150th anniversary of the Polish State Railways (PKP) service was held in Warsaw. Amidst much fanfare, a festive parade of nearly 20 steam locomotives ran back and forth in front of the spectators. The parade was followed by a nostalgic train that a steam locomotive hauled old passenger cars built many years ago from Warsaw to Glozick/Mazowiecka along the first section of the railway to provide passenger service in Poland. I also took this special train. A short time after leaving Warsaw, I looked out of the window of this train and was shocked to see France's TGV running next to our antique Polish train. It was a demonstration of French railway technology. My impression at the time was that they did it (Fig. 7.5).

In June 1997, Danish State Railways (DSB) fulfilled a long-held dream when it introduced a railway service across the Great Belt Straight. Up until this point, there had been no bridges or tunnels to connect Copenhagen on Sjælland Island with

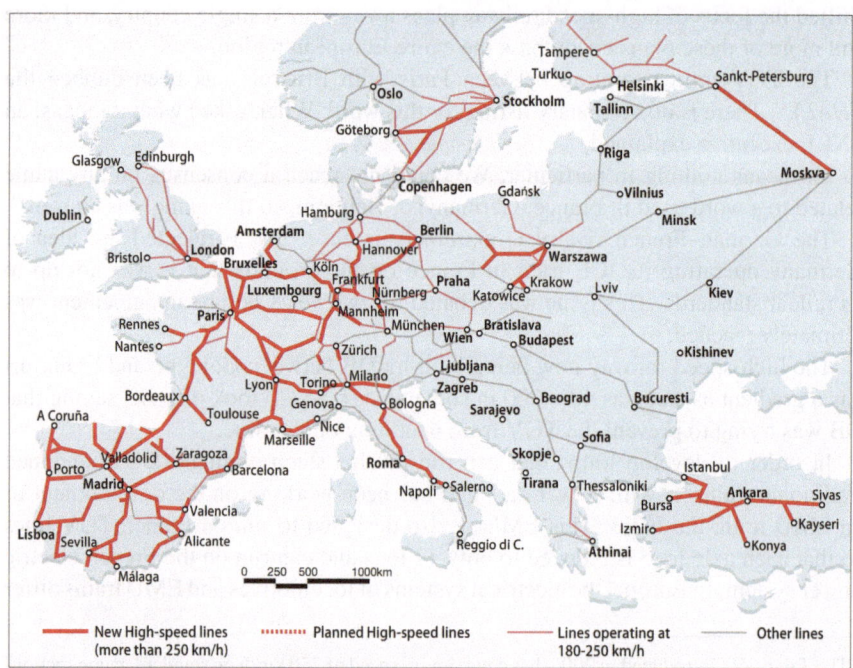

Fig. 7.5 European High-Speed Railway Network 2025. *Source* High-Speed Railways of the World JARTS 2014

Odense, the town in which the writer Hans Christian Anderson was born, on Fyn Island. Cars and trains had both been ferried across the straight. The completion of a seven-kilometer bridge and eight-kilometer undersea tunnel meant that passengers would now be able to reach Denmark's major cities by direct train without boarding a ferry. Not only this, but for the first time in history, Copenhagen would be linked directly with the railway network on the European continent.

Also at that time, DSB sponsored a railcar exhibition in Copenhagen. Intending to participate, a TGV trainset left France for the Danish border, only to be stopped at the German border. German customs officers would not allow the trainset to pass due to incomplete border customs documents or something like that. The TGV did not arrive in time to participate in the railcar exhibition, and rumors circulated that this might be German harassment.

France and Germany are not the only European countries involved in emotional conflicts. An English designer was hired to design the car bodies for the *Eurostar*, an international limited express running between Paris and London via the Channel Tunnel. A French designer, however, was in charge of the interior design. From what I understand, an argument ensued when UK insisted that it would not allow any train designed by a Frenchman into London's Waterloo Station.

Fig. 7.6 *Eurostar*, an international limited express linking Paris with London via Channel Tunnel. *Photo* provided by Tadashi Sumita

Waterloo Station takes its name from Waterloo, an old Belgium battlefield that was the site of the decisive battle between Napoleon and the British-allied forces. France, in turn, felt strongly that the English name *Eurostar* was inappropriate for an international train, but ultimately accepted the decision because the French company ALSTOM took the lead in receiving the order (Fig. 7.6).

It is not as if these European countries are constantly bickering. They do pull together at times on matters of overseas exports. On May 4, 1998, a special train comprising the latest French TGV double-decker passenger cars sandwiched between two German ICE power cars operated between Hannover and Göttingen in Germany. This train carried the German Minister of Economics, the heads of both French and German railway companies and rolling stock manufacturers, as well as the Chairman of the Taiwan High-speed Railway Business Entity. This was apparently a demonstration designed to appeal for cooperation between Germany and France.

In addition to railway lines built specifically for high-speed operations, some countries have achieved train speeds of 200 km/h and higher on railway lines built long ago.

In the UK, the IC225 limited express is capable of running at a maximum speed of 225 km/h from London to Edinburgh, Leeds and other cities. Sweden's tilting diesel *X2000* limited express and the U.S. *Metroliner*, which operates between New York and Washington D.C., both reach a maximum speed of 200 km/h. The U.S. is also working on a long-held dream of electrifying the line between New York and Boston, and its completion will bring a new type of train on to the railway scene.[8]

7.10   Weekly Round Trip for Russia's Signature Train

Although relatively unknown, Russia also operates a limited express EMU train at 200 km/h. In 1974, Soviet Railways announced that it had developed the ER200, an EMU limited express traveling at 200 km/h. This news was not followed by an announcement of specific operation plans, and the actual events that took place remain hazy. This train was apparently plagued by difficulties and was never readied for actual passenger service.

The Soviet Union was loathe to be outdone by France and Germany, but its technology, I imagine, was not up to the task. This story recalls similar competition over air travel. The Soviet Union built its own supersonic jet, an exact replica of the Concorde. It gained the nickname *Concordski*. In 1989, however, the Soviet jet crashed at the Paris Airshow.

Nevertheless, around 1984, there were reports that this train started running between Moscow and Leningrad (now Saint-Petersburg). The World Railway Conference was held in Moscow in 1989, and I received an invitation to serve as a panelist.

---

[8] The *Metroliner* is now discontinued, and the *Acela* Express began service between Boston—New York—Washington, D.C. in 2000. The maximum speed is 150 miles/h (240 km/h), though only on part of the route.

After the meeting was over, I asked if I could board this train and received the reply that I could.

I had done some research and was shocked to find that it is operated on only one round trip a week.[9] It simply departs from Moscow on Thursdays and again from Leningrad on Wednesdays. It is unusual for a country's signature train to merely make one round trip a week. It still seems strange, but perhaps this is due to technical problems, or the freight trains have a higher priority in the country's train schedule.

The Moscow-Saint Petersburg Line, the most important trunk line in Russia, also compares to Japan's Tokaido Line. The construction of this railway line, completed in 1851, was directed by George Whistler, an American railway engineer. The story goes that when Whistler expressed his opinion to Nicholas I, the Russian Emperor at that time, to give his opinion on the railway, the Emperor took out a ruler, placed it on a map, and drew a direct line between Moscow and Saint-Petersburg. "This," he said, "is the railway you will build."

The Russian railway employs broad gauge track. Though not as wide as Spain's, it does measure five feet (1,524 mm) across, 89 mm wider than the standard gauge used in other European countries. This is widely explained as the result of lessons learned by Russia from the Napoleon invasion. The truth is, apparently, that Russia adopted these wide tracks at the prompting of Whistler, a passionate advocate of broad-gauge tracks.

Despite the story of Nicholas I, this Russian railway line is not actually perfectly straight. There are curves with a radius of a thousand meters every so often to accommodate marshland and valley areas. Still, as a fairly straight stretch of track suitable for high-speed train running, it is a nineteenth century version of Italy's *Direttissima* line.

On the morning of this train ride, I received a telephone call at my hotel and was told that, for whatever reason, the train would be departing 30 minutes earlier than scheduled.

This came as a surprise. I know of trains departing later than scheduled, but an early departure, and by as much as 30 minutes, defies railway common sense. In Japan, a scene in which passengers arrived at the station in time for the scheduled departure only to find that the train had already left would spark a major outcry. But most shocking of all, the railway was telephoning their train passengers to let them know!

Figuring that the change in schedule would mean the train would be fairly empty, I went to Moscow's Leningrad Station with a guide from Intourist. The train was already waiting at the platform, and it was nearly full. For some reason, one of the eight cars was completely empty. I still have no idea how the railway managed to contact all of its passengers or why there were no passengers in that one car.

Once it departed and increased the speed, the train began to shake intensely. My back rubbed constantly against the seatback. I bought coffee and an open sandwich from the onboard vendor, and I tried to hold the coffee cup by keeping my back

---

[9] Currently, high-speed trains run between Moscow and Leningrad (now St. Petersburg) and operate 12 round trips per week at speeds of 250 km/h.

apart from the backrest of the seat, not to end up spilling the coffee. The landscape along this route was fairly monotonous—a long section of green forest interrupted occasionally by a town or large lake.

A large speedometer hung on the coach wall to show how fast we were traveling, but it did not move above 160 km/h for the first hour or two. Finally, the speedometer reached 200 km/h for a total of five minutes as we began to approach Leningrad. The train covered the 650 km section of the line in four hours, arriving in Leningrad pretty much as scheduled.

Chapter 8
Developing the World's Finest Railway

After the Tokaido *Shinkansen* began service, new *Shinkansen* lines were built in various directions in Japan. The author, involved in these projects, was faced with various challenges, especially in the case of the Tohoku and Joetsu *Shinkansen* lines into Tokyo, due to construction delays, and he prepared to submit a letter of resignation.

After the privatization of JNR, each railway company, reflecting its own circumstances, strove to create the world's best railway in its own way, including increase in speed and transportation capacity.

Meanwhile, France and other European countries also worked to increase speed in accordance with their national conditions. All of them were aiming to become the world's best railway. The author looks back on the evidence of how Japan and those countries' high-speed railways mutually stimulated each other and brought about technological progress and explains the differences in their respective policies, not from the viewpoint of a mere engineer but from that of an executive manager who served as chairman of East Japan Railway Company.

8.1 Struggling with the Tohoku *Shinkansen*

Around the time that France's TGV system was completed, Japan was nearing the point when service would be introduced on its Tohoku *Shinkansen*. In the spring of 1982, JNR finally announced the "Introduction of Service Between Omiya and Morioka".

I had been appointed Head of the North Tokyo Railway Administration Bureau the previous year and so was involved in preparing for the opening of the Tokyo end of the Tohoku *Shinkansen*. Little did I know at that time how many problems we were to struggle with before service would finally be introduced.

© Japan Railway Technical Service 2024
S. Yamanouchi, *If there were no Shinkansen*,
https://doi.org/10.1007/978-981-99-8890-7_8

First, there was the concern that we would not be able to meet the spring 1982 deadline. Purchasing the land on which to build the Tohoku *Shinkansen* had proved extremely difficult. This was especially true closer to Tokyo. It soon became obvious that there was no way the construction of the Tokyo-Omiya section would be completed in time, so we decided to scale our plans back. Service would initially be introduced only from Omiya.

Even so, we struggled for a long time over procurement of the land for tracks north of Omiya. By the time this finally went through, we were considerably behind schedule on our construction plans. Construction was unlikely to be complete by the end of May even with a final rush job.

This sparked a debate within JNR as to when exactly spring ended. Someone mentioned

Cherry blossoms[1] are not actually in bloom around Kushiro in Hokkaido until the middle of June,

Well, here we go. This *Shinkansen* line will eventually be extended all the way to Hokkaido, which means that, for this plan, spring is over at the end of June.

The Tohoku *Shinkansen* passenger service was introduced on June 23, 1982. Construction was still going on at Omiya Station when the opening ceremony was held there early that morning.

Before the service was introduced, though, an agreement on what to name each station had to be reached. Strange to say, the debate sparked by the name 'Nasushiobara' may actually have been the most difficult problem we faced in the Tohoku *Shinkansen* project.

Construction plans had used the temporary station name 'Shinnasu.' As we got closer to the official opening, however, certain factions began to insist that 'Shiobara' be incorporated into the name of the station. This became a point of fierce contention, with the local population divided into two camps. Ultimately, the debate even involved powerful politicians. The name issue—whether or not to include 'Shiobara'—left no room for compromise. In the end, we pushed the name 'Nasushiobara' through. The argument left such an unpleasant aftertaste, however, that I was forced to threaten the top executive with my resignation verbally.

All this fuss over the mere name of a railway station seemed ridiculous at the time. Perhaps, though, we should be grateful that the *Shinkansen* is seen as so valuable and generates such high expectations that people are willing to go to battle over these issues.

The issue involved in the Tohoku *Shinkansen* service that concerned me most was how passengers arriving in Omiya would be transported to Tokyo. The Keihin Tohoku Line EMU trains make the trip between Tokyo and Omiya in 50 minutes, but these trains could not accommodate *Shinkansen* passengers loaded down with luggage. They would at least need to be guaranteed a seat between Ueno and Omiya if they were to put up with the inconvenience of having to change trains.

To solve this problem, the '*Shinkansen* Relay Train,' a limited express that would shuttle connecting passengers back and forth this section, was introduced (Fig. 8.1).

---

[1] Cherry blossoms are the harbinger of spring in Japan.

Fig. 8.1 *Shinkansen Relay Train.* This intermediary limited express carried Tohoku *Shinkansen* passengers between Omiya and Ueno as a provisional measure until the entire line was completed. *Photo* provided by the Railway Museum

Even this solution, however, generated quite a few formidable questions. What timeframe would we allow for connecting passengers at Omiya Station? What would we do in the unlikely event that passengers missed this train? Would the Relay Train accommodate all of the passengers from an 800-seat *Shinkansen* train? What was the contingency plan for a delayed Relay Train? And so on and so forth.

Three years until the Tohoku *Shinkansen* was extended at last to Ueno Station had no serious problems that we had worried about.

When service was first introduced, the Tohoku *Shinkansen* provided the ultimate in passenger comfort. It was reputed to be so little vibrating that a single cigarette propped up against the window during a test run did not fall over, and I tested their claim. My cigarette did, in fact, stand on its end for as long as a minute. Although the quality has now fallen off somewhat, the same could have been said for the Nagano *Shinkansen* when it was first opened.

8.2 The Emperor and Empress Journey to Nasu

In the summer, we were honored to have the Emperor and Empress of Japan onboard the Tohoku *Shinkansen* for their journey to the Nasu Summer Palace. Normally, their majesties board the royal train at the Royal platform, reserved especially for their use at Harajuku Station, and travel directly to Kuroiso Station. To mark the Tohoku

Line opening, however, they honored us by transferring to the *Shinkansen* at Omiya Station.

With the exception of journeys for official state functions, it is a rule that the Emperor is guided by the Head of the Railway Administration Bureau. Once the Emperor and Empress had disembarked from the royal train at the Omiya Station, the station master and I were to escort their majesties. But the transfer would involve a much longer distance than at Harajuku Station. In order to properly carry out our roles as guides, we would have to slow our pace to match that of the Emperor and Empress. Since we would be unable to look directly at their majesties, we arranged for assistant station masters to stand at a number of strategic points and signal to us to increase or slow our pace.

I was concerned about another matter. The prefectural governor and his office staff were to meet the Emperor and Empress on the platform of Nasushiobara Station, and I was to chat with their majesties while we waited on the platform for them to line up to see the majesties off in front of the station.

I had accompanied the Emperor on nearly 60 train journeys, including riding in the driver's cab as the executive in charge of the royal train's operations when the director of the EMU Train Division. This, however, would be the first time that I would be conversing with his majesty. I had no idea what to speak with him about.

After thinking it over, I decided to describe the view from the *Shinkansen* platform since I remembered hearing that the Emperor often asked the names of the mountains and rivers he saw as he passed in the royal train.

I told the station master that I would come to Nasushiobara to familiarize myself on a day when the Nasu mountain range was in clear view and asked him to call me when the weather was good.

I waited a week, then two, but received no telephone call. I finally called him myself, but was told that there was very low visibility. On the day before the royal train was to run, I was struggling to find another topic of conversation when I received a call from the station master.

Head, the weather's finally cleared up.

I immediately jumped on a *Shinkansen* bound for Nasushiobara Station.

The following day, the weather was beautifully clear, and full of confidence, I described the mountains we were looking at.

The Emperor nodded as he listened to my explanation, but relaxing in the station master's office later, I realized with a jolt of embarrassment that, having made the journey to Nasu for many years, his majesty was in a much better position to know the names of these mountains.

To return to our struggle with *Shinkansen* development, the permanent terminal station on the Tohoku Line was to be Tokyo Station. Omiya was nothing more than a temporary solution. To end the inconvenience of having to change trains at Omiya Station, the Tokyo-Omiya section would have to be completed soon. The Construction Department was making an enormous effort, but there were no plans yet for construction near Kanda. We had not been able to purchase the necessary land.

To bypass the land problem, the original plans were modified. A *Shinkansen* station was to be built beneath Ueno Station to serve as the terminal station in Tokyo on the Tohoku Line. On March 14, 1985, the Tohoku *Shinkansen* finally arrived in the heart of Tokyo.

The construction of Ueno Terminal, however, brought its own share of problems.

The terminal was to be built 33 m below ground. Tokyo Station's platforms for the Sobu and Yokosuka Lines had been built underground, which gave us some experience with underground stations. This, however, was to be the first *Shinkansen* station to be built below ground. Most passengers arriving on the Yamanote and Keihin Tohoku Lines at Ueno Station would go upstairs to a transfer concourse before taking a long escalator to descend the 40 m below ground to the *Shinkansen* platform (Fig. 8.2).

Though our problems were behind us, I thought that the Tohoku *Shinkansen* was finally complete. Things were not quite this simple, though. Passengers arriving from the Tohoku and Niigata continued to complain that they were supposed to be arriving in Tokyo, which to them meant Tokyo Station not Ueno Station.

Work to extend the *Shinkansen* to Tokyo Station was restarted after JNR was privatized, and the original plans were finally completed on June 20, 1991. The construction of this Tokyo-Ueno section cost 128.2 billion yen. Although the different

Fig. 8.2 Passengers take the long Ueno station escalator to the *Shinkansen* concourse. *Photo* provided by Tetsuro Aikawa

decades of construction made comparisons difficult, the mere 3.6 km section between Tokyo and Ueno cost a third of the budget for the entire Tokaido *Shinkansen*.

Not only had we faced numerous difficulties in extending the *Shinkansen* to Tokyo, but neither Tohoku nor Joetsu *Shinkansen's* construction works went smoothly. These high-speed lines were plagued by problems from beginning to end.

8.3   Tohoku and Joetsu *Shinkansen* Fall below Expectations

The number of Tohoku and Joetsu *Shinkansen* passengers fell far below the original projections.

Development plans had estimated 90,000 passengers daily on the Tohoku *Shinkansen* and 70,000 passengers on the Joetsu *Shinkansen* by the tenth year of service. In reality, passengers numbered only 57,000 and 41,000 on the Tohoku and Joetsu Lines, respectively. The original projections proved overestimated.

This was the reverse of our experience with a higher number of passengers on the Tokaido *Shinkansen* than had been predicted. The factors behind this were the downturn from the rapid economic growth in Japan and the six-year delay in introducing passenger service.

There were also a number of points on the finished *Shinkansen* for which JNR must take the blame. If the Tokaido *Shinkansen* symbolizes a passionate and energetic JNR, then the Tohoku and Joetsu Lines embody a stumbling railway struggling with financial and labor problems.

First, construction took 20 years to complete, four times the original projection. Second, the cost of the Tohoku *Shinkansen* construction alone totaled 2.7 trillion yen. Though the times are different, this is eight times the budget for the Tokaido *Shinkansen*, which is nearly the same length as the Tohoku Line. Although the Tokaido *Shinkansen* has three times the traffic density of the Tohoku *Shinkansen* and is five times as the Joetsu *Shinkansen*, the per-kilometer construction cost of these *Shinkansen* was seven to eight times that of the Tokaido *Shinkansen*. Even when adjusted for inflation, this is a considerable increase.

The high-quality facilities and EMUs on these *Shinkansen* do cost money. There were fewer problems than we had experienced with the Tokaido *Shinkansen*, as well. It was considered that the number of tunnels these regions would require, stringent measures were taken to reduce vibration along the lines, and the snowy Tohoku and Joetsu regions required measures to prevent snow damage to operate these trains with no problems. We applied every bitter lesson we had learned from building the Tokaido *Shinkansen* to the development of these two lines. To ensure that the lines stood up to rain, the tracks were laid on concrete structures rather than on embankments.

But did we really make the right decision?

When plans to construct the Tokaido *Shinkansen* were announced, politicians and the public could not identify the significance of a high-speed railway, and many of the opinions were critical. For this reason, the Tokaido *Shinkansen* started with an unreasonably low construction cost estimate which led to major problems later when the actual cost exceeded these estimates by a considerable margin. Still, JNR made a real effort to build the *Shinkansen* at the lowest possible cost.

However, when the Tohoku and Joetsu *Shinkansen* were built, there was no such opposition working to control costs. Public opinion was essentially in favor of the plan. Politicians and local residents were ardently in favor of the construction of a *Shinkansen* close to home. The only opposition was on environmental issues, which finally led to the cancelation of the Narita *Shinkansen*. But this only served to increase the money spent on other *Shinkansen*.

Encouraged by the success of the Tokaido *Shinkansen*, the JNR Passenger and Operations Departments reversed their earlier position to advocate for the *Shinkansen* and push for even more luxurious facilities. The quality of station facilities, in particular, was improved. Based on its experience with the Tokaido *Shinkansen* troubles, the Maintenance Department requested robust equipment with fewer breakdowns.

The reality of unsatisfactory maintenance work due to labour problems in the fields was also part of the background to this. Once construction began, in the station section, for example, the conventional line facilities inevitably had to be modified dramatically. The fact that the *Shinkansen* would be passing through these stations provided the perfect opportunity to renovate, and certain officials even went out of their way to develop plans that would replace old facilities with new ones as much as possible.

At the time the Tokaido *Shinkansen* was built, the Construction Department had worked hard to reduce construction costs and maintained a strong hold on all wasteful spending. Now, however, even this department was less motivated and less powerful. Attempts made to cut construction costs had little effect on a JNR deficit that was snowballing out of control.

There is a strong impression that the construction work proceeded with a high-cost structure under the collusive tendency in the name of estimated construction prices and nominated competitive bidding.

Private companies are motivated by profitability. The goal of profitability means building more modest facilities for railway projects with lower traffic demand. In line with this fundamental business principle, the facilities developed for the Tohoku and Joetsu *Shinkansen* should have been more modest than those on the Tokaido *Shinkansen*. The opposite, in fact, actually occurred. New projects were constantly pursued, and grander facilities and more costly new technologies were chosen over more modest ones. The principle of profitability was secondary. This illustrates an inherent weakness in government-owned companies.

Despite their grand facilities, the Tohoku and Joetsu *Shinkansen* technology is nothing to boast about. These trains do not run faster than 210 km/h on the Tokaido *Shinkansen*, and no obvious changes were made to the EMUs. ATC devices and other electric equipment also remain essentially the same.

A reporter from a French railway magazine described his impressions after taking a test ride soon after the Tohoku *Shinkansen* had been completed.

> To be honest, the EMUs on the Tohoku and Joetsu *Shinkansen* are essentially the same as those on the Tokaido and Sanyo *Shinkansen*. The strongest impression on this reporter was that the color of the cars is slightly different. The front of the trains is equipped with

snowplows; the bottoms of the car bodies are covered; and there are plasma displays in the driver's cab. These, and the fact that the motor car is slightly more powerful, seem to be the only new features. The Japanese people are proud of the fact that, with the completion of these lines, they can now boast of two more *Shinkansen*. But these new *Shinkansens* are replicas of the Tokaido *Shinkansen*, a concept developed 20 years ago. Is conservatism the way that the Country of the Rising Sun will pursue?

Changes were, in fact, made. Hoping to increase train speeds to 260 km/h, JNR developed the 951 and 961 in the 1970s and tested these two high-speed prototype EMU trains on the Sanyo *Shinkansen*. The maximum speed achieved by the 951 during its test run was only 286 km/h, a speed that does not even surpass the record set by France in 1955. Although the failure to achieve higher speeds was due in part to technical problems with the prototypes, labor problems also had a significant effect.

Several rounds of negotiations with the labor unions were needed to conduct even a single test run. The unions would bring up safety issues and other problems and then push for solutions to their numerous demands. Under these conditions, it was impossible to conduct these types of tests. The 961 was an improvement over the 951 with all of its technical problems. This prototype was built to test the technology to be used in the design of new EMU trains for the Tohoku and Joetsu *Shinkansen*. The prototype was abandoned, however, without conducting satisfactory test runs.

Despite these failures, the Tohoku and Joetsu *Shinkansen* are different from the Tokaido *Shinkansen* in ways that are not immediately obvious. First, the train schedule was modified. *Yamabiko* and *Asahi* super express trains stop at fewer stations than the *Aoba* and *Toki* trains which stop at every station. This pattern is the same as *Hikari* and *Kodama* of the Tokaido *Shinkansen*, but super express trains stop at considerably more stations than them.

The terminal station is in Morioka, a city that is quite a bit smaller than Kyoto or Osaka; the only large city on the Tohoku *Shinkansen* is Sendai. The super express stops at such mid-size cities as Utsunomiya, Koriyama, Fukushima, Takasaki, and Nagaoka to mitigate the decline in passenger numbers. When service was first introduced, in fact, every train stopped at all the stations between Sendai and Morioka. With only one super express train departing hourly, we had no choice but to increase the number of station stops. This in itself is a significant change in the *Shinkansen* scheme.

The primary purpose of the Tokaido *Shinkansen* was to create new railway lines in order to increase regional transport capacity and offset existing Tokaido Lines that were already stretched to their limits. Similar conditions existed between Tokyo and Sendai and Tokyo and Niigata. This was not true for the main Tohoku Line north of Sendai. The original purpose of increasing transport capacity through *Shinkansen* lines had shifted to an increase only in speed. Fewer passengers on these lines, however, meant that the train frequency was reduced, and they stopped at every station on the line, which contradicted the *Shinkansen* purpose of increasing speed.

Similar problems awaited the Joetsu *Shinkansen*. With no major cities along this line, the simplified *Hikari-type* train schedule was abandoned. Instead, a variety of *Asahi* super express trains stopping at different intermediate stations were scheduled on the Joetsu *Shinkansen*. This was the first *Shinkansen* service not to run on a pattern timetable under which a certain type of train departed at a specified time each hour.

Another substantial change involved the number of onboard staff. When the service of Tokaido *Shinkansen* was first introduced, three drivers and an inspector in the driver's cab, and five conductors were on board. On the Tohoku and Joetsu *Shinkansen*, the numbers were reduced to one driver and two conductors. This was one way of rationalizing operations. This cut in staff prompted the installation of the plasma display in the driver's cab. This display indicates any EMU abnormality and allows onboard staff to make simple repairs by remote control.

Thereby it eliminated the need for the second driver, whose job was to inspect for abnormalities and make simple repairs. The driver's cab on the Tohoku and Joetsu EMU trains was, I believe, the first to be equipped with this display. Although construction costs were high for these two *Shinkansen* lines, these innovations have reduced operating costs. This does represent an advancement, albeit a modest one. The TGV, however, is not equipped with a plasma display and yet has always been operated by only one driver.

French engineers developed the TGV by thoroughly studying Japan's *Shinkansen* technology, avoiding its use as far as possible, and working to develop the kind of technology they believed to be theoretically sound and test-driving trains for seven million kilometers. By contrast, Japanese engineers developed the *Shinkansen* by aggressively adopting and rearranging proven European railway technology. This illustrates what may be the essential difference between these two countries' cultures of technology, while I believe that Japan made the right decision in basing the *Shinkansen* on the philosophy it did.

Very little progress was made over the following 20 years. By the time that the TGV began carrying passengers, Japan's system was, technically speaking, a step or two behind the French.

8.4 *Nozomi* and *MAX* Debut the First Full-Model Change

The debut of JR Central's series 300 *Nozomi* in 1992 represented the first full-model change to *Shinkansen* EMUs. This EMU reached a maximum speed of only 270 km/ h, but this was due to the many relatively sharp curves on the Tokaido *Shinkansen* rather than poor EMU performance.

Designed for speeds of 210 km/h, the Tokaido *Shinkansen* incorporated a number of curves with a radius of 2,500 m. The Sanyo, Tohoku, and Joetsu *Shinkansen* were designed for a maximum speed of 260 km/h with relatively gentle curves with a radius of 4,000 m or more to enable trains to run at higher speeds. The only exceptions were the section between Tokyo and Omiya and other special sections. France's TGV lines are also constructed with curves with a radius of 4,000 m or more.

The *Nozomi* EMU trains adopted full-scale electronic technology in their control systems and bogie structure of a completely different design from that used in past

Shinkansen EMU trains. This marked the first real advance in EMU design in the 28 years since the introduction of the Tokaido *Shinkansen* service.

JR East then began work on its own advanced prototype EMU train, the *STAR 21*, to make next-generation designs for *Shinkansen* EMUs. This prototype was equipped with a variety of new technologies to enable engineers to resolve a diverse set of technical questions: could these technologies really be used in passenger service trains? Which technologies were superior? How far could the performance be exploited? To what extent could trains be lightened?

JR East's first Chairman, Isamu Yamashita, ordered engineers to cut train weight by half. Although engineers balked at this idea at first, they ultimately accomplished this goal.

The Chairman explained, "There would have been no conceptual innovation if I had merely told engineers to reduce the weight of EMUs by, say 30%. By ordering a 50% reduction, engineers were able to break out of a rigid traditional mindset."

Three types of EMUs were developed, each with a unique car body structure. Three different bogie structures, including both the typical double bogie car and the articulated car used on the French TGV trains, were also tested. The prototypes featured various covers for pantographs and noise reduction, and the head shape is also designed with a completely different shape in the front and back of the trainset.

These prototypes were test run for 210,000 km over five years, providing us with valuable data. This data indicated no significant difference between the double bogie and articulated type of EMU in terms of either operational stability or passenger comfort (Fig. 8.3).

Fig. 8.3 The *STAR 21* test train challenged maximum speeds. *Photo* provided by Kotsu Shinbunsha

Fig. 8.4 Driver's seat at the moment when the *STAR 21* set a new speed record of 425 km/h between Nagaoka and Niigata on September 9, 1993. *Photo* provided by Kotsu Shinbunsha

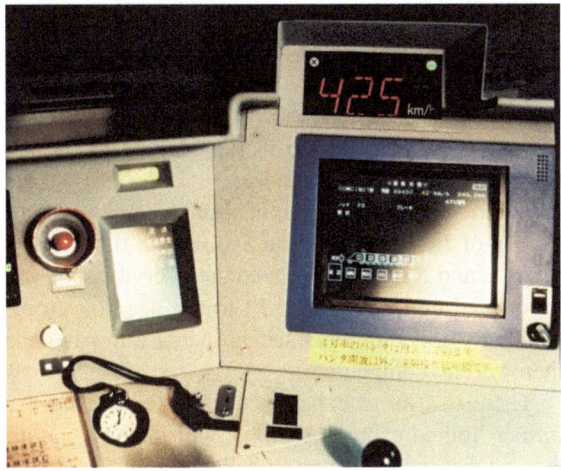

The *STAR 21* was used to test how fast it could run, with speed gradually being increased during a series of high-speed runs between Nagaoka and Niigata in the autumn of 1993.

On September 9, JR East set a new speed record of 363.6 km/h. The company subsequently broke its own record on December 21 when the prototype reached a new record speed of 425 km/h. Considering the previous speed record of 350 km/h for a Japanese *Shinkansen* EMU train held by the JR West high-speed prototype train, *WIN 350*, the record set during this test run was revolutionary for Japan's railway system (Fig. 8.4).

JR East's record was broken three years later when the JR Central prototype '300X' reached a maximum speed of 443 km/h.

The particular rivalry generated by the privatization and division of JNR provided a significant push in terms of technological progress. This idea has been rejected on the grounds that the regional division of JNR does not bring about such a sense of rivalry between JR Group companies. This, however, is not true at all. There has been quite fierce rivalry with regard to technology, service, business development, and even, at times, commodity sales. Although this did occasionally threaten to get out of hand, rivalry itself is not a negative force.

After her first visit to Japan, Ms. Jolene M. Molitoris, Federal Railroad Administrator, U.S. Department of Transportation, wrote me a letter in which she described her impression of our railway system. "Even though the JR Group companies do not compete in the same market, they do try to outdo one another in terms of excellence."

Around the time that these speed experiments were coming to an end, JR East's stance on increasing train speed began to change. High-level executives began to question how important it was to increase train speed further. To be honest, this stance left me somewhat perplexed. Speeding up is, after all, an engineer's dream, and faster trains give a positive image of the company's technical level.

We should look, however, at how much Japan's pioneering work with the Tokaido *Shinkansen* in the world of high-speed trains did to enhance the image of Japanese railways. Most JNR employees certainly took pride in this accomplishment, but we should also admit that we overstated the case when we called Japan's railway technology the best in the world. Making the *Shinkansen* system more luxurious might also have been a mistake.

In terms of increasing train speeds, however, there are no airlines parallel to the section of JR East where the *Shinkansen* runs. Flights between Tokyo and Sendai, Niigata, and Hanamaki stopped their service soon after the *Shinkansen* service was introduced on these routes. I had no idea how to respond when questioned about whether we would really see an increase in passengers with a faster service when there was no competition.

There is a lot of experience that speeding up the express trains, including the former limited express *Kodama*, will increase the number of passengers. At least, this is what we had come to believe. When questioned on this point, however, my confidence waned.

Increasing the speed of trains operating north of Omiya on the Tohoku *Shinkansen* to 300 km/h meant that passengers would reach Sendai in one hour and 28 minutes and Morioka in about two hours and 13 minutes. At a maximum speed of 275 km/h and no intermediate stops, *Yamabiko* trains reach Sendai from Tokyo in one hour and 36 minutes; it only saves eight minutes. I couldn't honestly insist that an eight-minute difference would increase the number of passengers on this line.

On top of this, trains would undoubtedly consume more energy, and the measures required to reduce noise would require spending more money.

In 1997, the new E2 and E3 EMU train series made their debut on the Tohoku and Nagano *Shinkansen*, operating at a maximum speed of 275 km/h. We decided, however, to hold off before increasing the speed of our trains any further (Fig. 8.5).

Later, we began to give more priority to passengers to sit down than to building faster trains. *Shinkansen* fares and surcharges are costly. With the exception of the year end, the beginning of the year, Bon holidays, and other particularly busy times of the year, passengers rightly expect to be able to sit on the train as part of the price of the ticket. Recently, however, an increasing number of passengers are using *Shinkansen* trains to commute to work, and more and more passengers have been forced to stand between Oyama and Kumagaya. To alleviate their discomfort, we decided to develop a *Shinkansen* EMU train with as many seats as possible.

In 1994, the *MAX* E1 series, composed entirely of double-decker EMUs, made its debut. Formed of 12 EMUs, this train provides 1,135 seats. The *MAX* E4 series, a similar double-decker EMU train with 16 cars accommodating 1,634 passengers, was introduced in 1997. The *MAX* E4 features the largest transport capacity in the world (Fig. 8.6).[2]

By developing the E1 and E4 series, we have also cut the number of cars equipped with electric motors to half. Our *STAR 21* prototype gave us the technology to build ultra-light car bodies and increase the electric motor power output, as well as improve

---

[2] The E4 is no longer in service.

Fig. 8.5 *Asama*, the new E2 series, Nagano *Shinkansen* EMU train. *Photo* provided by the Railway Museum

Fig. 8.6 *MAX* E4 series, the latest lightweight double-decker *Shinkansen* train. *Photo* provided by the Railway Museum

the electronics technology to provide extremely precise control. The E1 and E4 series are perhaps 'the most economical high-speed EMU trains in the world.'

8.5   Individual JR Group Companies Started Technical Development

Since the privatization of JNR, JR companies have been focusing on technological development in order to catch up with the technological development that lagged behind during the JNR era. Significant progress was made in *Shinkansen* technology over the subsequent ten years. The division of JNR has given each of JR companies their own unique technological development. JR Central and JR West were passionate about increasing train speed. Unlike JR East, subject to intense competition from air flights, they naturally stressed speeding up. In 1997, JR West's focus on speed produced the *Nozomi*[3]; with its maximum speed of 300 km/h, this EMU train traveled between Shin-Osaka and Hakata in a mere two and half hours. At the time of privatization, the same distance took two hours and 59 minutes to cover. In only ten years, the trip time had been reduced by almost 30 minutes.

In terms of developing technology, JR East gradually shifted away from speed and began to focus on enhancing economic efficiency, safety, and other basic aspects of service. These differing points of view on technological development are also, I believe, the result of JNR privatization and breakup.

Only three types of *Shinkansen* EMUs were operating at the time of privatization, and the basic design of each was essentially identical to the first Tokaido *Shinkansen* EMUs. Japan's railway system now has ten different types of EMUs that are so much more technologically advanced that they have little to do with their predecessors.

Some people say that the division of JNR may have led to a dispersal of technical capacity and a decline in technological development. This, I believe, is far from true. JR Group companies do indeed develop different types of EMUs, but this is natural in light of the various management conditions and markets they deal with. Private railways also operate different types of EMU trains, but this does not mean that the technology one uses is inferior to that of another. Rather, it is precisely this competition and diversification that accelerates progress in railway technology.

Besides, the same manufacturers produce the EMUs for all JR companies. The diversification of EMUs has certainly complicated matters for EMU designers and manufacturers, but the basic technology to develop the cars is expected to be properly passed on. The bulk of the technology developed for one JR company is incorporated into the products of other JR companies. As is the case with automobiles and home electronics, the continuous composition between many companies for new product features and designs accelerates technological progress. The restructuring of JNR has actually made a significant contribution to innovations in Japanese railway technology.

---

[3] This train is operated with the Series 500.

So, is the European railway technology exhibited in France's TGV and Germany's ICE or the Japanese technology of the *Shinkansen* superior? Although technology is difficult to compare, I do not see any significant difference, and not simply because the Japanese *Shinkansen* now runs at 300 km/h.

Maximum speed is one indicator of technical standards, but this matter is not quite so simple. Who is to say whether the technology behind the Concorde or the jumbo jet is superior? Whether the technology behind a Toyota or a car like the Porsche or Ferrari is more advanced?

The important question is not whether a train with power cars or an EMU train is superior, either. The Japanese *Shinkansen* has a greater transportation capacity and less costly EMUs on a per-seat basis, but these prove meaningless in terms of comparing technologies. The *Shinkansen* system was designed with large-scale EMUs because a high-speed train in Japan must transport a large number of passengers, and it was understood from the outset that these trains would not operate on conventional lines. In Europe, however, the relatively small number of rail passengers and high-speed trains running through to conventional lines resulted in more compact car bodies. Rather than a difference in technological capability, this is a difference in need.

This topic warrants more detail.

I have mentioned that each railway sector saw significant technological innovations immediately before the Tokaido *Shinkansen* was constructed. Similarly, the ground was being laid for another substantial innovation in railway technology in the 1980s. Revolutionary changes were taking place in the application of electronics technology, particularly with regard to the electric motor at the heart of high-speed EMU trains. Despite the long struggle by railcar engineers to develop alternating current electrification, the three-phase alternating current electric motors used in ordinary factories and houses proved difficult to apply to railway technology, and direct current electric motors were necessary to be used.

8.6 Fewer Motors and Pantographs

EMU trains and locomotives required rectifiers to convert alternating current to direct current, and inspecting and repairing direct current electric motors was time-consuming. Electronic technology resolved these issues by allowing railways to use three-phase alternating current electric motors. By the end of the 1970s, this technology was nearly ready, and Europe began adopting it on a trial basis for its locomotives. In Japan, certain private railways and streetcar operators began to adopt these motors, as well.

JR Central *Nozomi* was the first Japanese *Shinkansen* EMU to make full use of this technology. Following a series of tests with the *STAR 21* prototype EMU, JR East first adopted this method for its *MAX* E1 double-decker *Shinkansen* EMU train. Although TGV originally ran on direct current electric motors, France began to replace these with alternating current motors for its Atlantic line. German ICE has been electronically controlled since its inception. The application of the three-phase alternating current motor has dramatically reduced the weight of the electric motors used on trains.

Italian company Fiat plans to use motors manufactured by Hitachi on its new high-speed tilting train. Planned for completion by 2002, this train will operate at 300 km/h. The electric motors manufactured in Japan are the lightest and most powerful in the world.

We were also able to manufacture lighter car bodies. Another advance in design technology has been made possible through the use of aluminum, stainless steel, plastic, and other new materials for car bodies. An aluminum car body and compact, lightweight alternating current electric motors alone cut the weight of the *STAR 21* cut by half. Comparison testing on this aluminum car body was conducted with three different types of construction—honeycomb, extrusion, and riveted construction as used in airplanes. The results of these JR East tests were shared with all JR group companies.

To reduce the cost of the *MAX* E1 series, we constructed car bodies of double-decker EMUs with common steel, which would have been impossible before the advent of this electronic technology. Despite the increase in height, these double-decker EMUs are lighter than the original *Shinkansen* EMU trains.

One key to these lightweight EMU trains is that the powerful alternating current motors reduce the number of motors needed on each train. Every EMU on the first trains manufactured for the Tokaido, Tohoku, and Joetsu *Shinkansen* was equipped with electric motors. Only six of the 12 cars forming the *MAX* E1 series trainsets are equipped with motors.

Aluminum car bodies have further reduced the weight of the *MAX* E4 series EMU trains. These EMUs weigh 2.1 tons per meter; the *Asama* E2 series is even lighter at 1.7 tons per meter. Our old 200 series weighed in at 2.3 tons. The original TGV car bodies, which are quite a bit smaller than the cars on *Shinkansen* EMU trains, were relatively heavy at 1.9 tons per meter. Japanese *Shinkansen* car body weight standards have been achieved, however, with the most recent double-decker TGV train, the *DUPLEX* weighs about the same as the E4 series. Compared to the Japanese *Shinkansen*, it is comparable in terms of the lightness of the car body.

This reduction in weight is extremely significant in high-speed EMU trains. Not only are lighter trains more stable at higher speeds, but they also subject the tracks to less load and consume much less energy. The reduction in weight also reduces noise to a certain extent and changes tone quality.

These advancements have obviously improved passenger comfort on *Shinkansen* EMU trains. SNCF executives were impressed by how little the *Asama* EMU trains swayed. The initial Tokaido *Shinkansen* bogies were nearly replicas of standard German bogies, but the bogies of the latest *Shinkansen* EMU trains have become a uniquely Japanese design.[4]

---

[4] The appearance of this bogie is similar to that of the 'Minden German bogie,' with the front and rear sides of the axle box supported by plate springs, but the plate spring supports on the bogie side are connected via rubber, and the elasticity of the rubber can be adjusted to select the optimum

We have also reduced the number of pantographs used on our high-speed EMU trains. A 12car trainset originally ran with six pantographs; now, only two are used. French TGV has always run with a single pantograph. This represents a philosophical difference more than a difference in technical standards.

Since Sanyo *Shinkansen* was built, Japanese *Shinkansen* tracks have been laid directly on a foundation of concrete. Replacing the ballast used for so many years this method produces a more stable track, thereby increasing safety and reducing repair costs. The construction costs for this method, however, are much higher.

French TGV and other high-speed trains, on the other hand, continue to be built on a traditional ballasted track base. In general, there are fewer problems with ballast construction in Europe, where the ground is firmer than in a country like Japan, which is subject to considerable rainfall and earthquakes. These conditions also help keep construction costs down. Again, this is not an issue of technological superiority.

There are no longer significant differences between the Japanese and European overhead contact wire structures, either. The original overhead contact wire structure for the high-speed running of the Tokaido *Shinkansen* was fairly complicated, but this has since been replaced by the simple structure used by French TGV and other European high-speed trains.

There has also been progress made in terms of computerization. Japan pioneered seat reservation systems on the Tokaido *Shinkansen* and later introduced COMTRAC, a Computer Aided Traffic Control system. European high-speed railways also employ similar systems. In 1996, however, JR East introduced a revolutionary new traffic control system, COSMOS, which fundamentally changed the concepts and techniques of traditional traffic control systems. In this regard, Japanese technology is obviously a step ahead. The technology developed by Japan's powerful computer companies made this system possible.

European railways, on the other hand, are focusing on the development of a new 21st-century wireless ATC system called 'ERTMS.' This system has the potential to revolutionize our traditional signal system with its long history of passing a current through the rails. As with every sector, the railway community is also experiencing a radical technological revolution based on electronics and information technology. Companies that fail to pursue technological innovation will suddenly find themselves left behind, as we experienced at JNR.

When comparing Japanese and European railway technology, the expense of the *Shinkansen* is a point that I consistently return to. Despite the negligible difference in operating and EMU costs, construction costs for *Shinkansen* ground facilities are considerably higher than those of their European counterparts.

Construction costs for the Tohoku and Joetsu *Shinkansen* reached five to six billion yen per kilometer. This is four times the cost estimated in the original plans. By comparison, construction of the Paris-Lyon section of France's TGV line was completed around the same time at a cost of approximately 500 million yen per kilometer.

---

value for the axle box support rigidity in the lateral and longitudinal directions, which is a unique device that dramatically improves running stability.

We should be clear that Tohoku and Joetsu construction took much longer than planned, and Japan was hit by skyrocketing prices resulting from the oil crisis over this extended period. Price levels from the time that the Tokaido *Shinkansen* was constructed show the wholesale price index and the construction material price index to be 2.4 to 2.6 times higher when the Tohoku and Joetsu lines were completed. With construction costs for the Tokaido *Shinkansen* at approximately 640 million yen per kilometer, construction with a similar cost structure would have brought the Tohoku and Joetsu *Shinkansen* in at 1.3 to 1.5 billion yen per kilometer, even with inflation. The reality is that these lines cost four times this amount.

The Tohoku and Joetsu lines required a considerable number of long tunnels. Extremely high-quality facilities were also built, and various measures have been taken to deal with snow. Still, those who feel that all of this does not justify the considerable expense involved are not wrong. The construction of the Tohoku and Joetsu *Shinkansen* cost an enormous 3.3 trillion yen, and considering the capital cost burden, these two *Shinkansen* lines are quite unprofitable.[5] Even figuring for inflation, they could have been built for little more than one trillion yen if the same care had been taken with these lines as with the construction of the Tokaido *Shinkansen*. If this had been the case, they would now be turning a solid profit.

Nagano *Shinkansen*, for example, which began carrying passengers in 1997, was constructed at approximately seven billion yen per kilometer. The north TGV line in France was completed in 1993 for approximately one billion yen per kilometer. The German high-speed train line, as well, was built for 4.5 billion yen per kilometer.

---

[5] JR East's Shinkansen has a 37% profit after depreciation in its financial results for March 2019. This is because the number of passengers has increased by approximately 1.5 times compared to the original forecast, the work efficiency has doubled due to rationalization, and the long-term interest rate has decreased from 6.2% to 1.9%.

Chapter 9
Privatization as a Cultural Revolution

The *Shinkansen* was a revolution where Japanese railways impacted the world of railways. Another revolution where Japanese railways impacted them was the privatization of the railway industry.

Privatization has been a greater success than anticipated. This chapter introduces various challenges that JR has undertaken since privatization, such as technical innovation of rolling stock, provision of unconventional station services, and technological improvement. The author established the Charter of Railway Business and promoted these challenges, and we hope readers will enjoy his honest and frank explanations.

He will also share his own experiences of management reforms implemented by European countries under the influence of Japanese privatization.

9.1   Privatization and Division of JNR in Two Years

By the end of its reign, JNR was catastrophic. It faced a total debt of 25.1 trillion yen. This was nearly twice Brazil's debt at the time, clearly making JNR the poorest-performing company in the world.

JNR's revenue was 3.6 trillion yen in 1986, but 1.3 trillion of this evaporated in interest payments.

Relations with labor unions were also tense. Labor unions insisted on getting their approval for the smallest change to assigned tasks and every new experiment, which stymied the rationalization of our operating systems.

Just before the privatization of JNR, a sense of crisis led to a reduction in personnel. Still, the company employed a staff of 420,000 in 1979, and personnel expenses accounted for 66% of revenue. Despite these conditions, JNR continued to make one trillion yen's worth of capital investment every year. JNR was a stereotypical government-run company, operating contrary to the principles of private business.

© Japan Railway Technical Service 2024
S. Yamanouchi, *If there were no Shinkansen*,
https://doi.org/10.1007/978-981-99-8890-7_9

In 1982, the Japanese government's Special Administration Survey Committee outlined the course of action for the privatization and division of JNR, but JNR was reluctant to follow this policy. In 1985, the entire management, including the president, was replaced and began to prepare actively for privatization and breakup. Only two years until privatization in April 1987 remained, but there was still a mountain of work to do. For setting up the new JR companies, the new organization had to be structured, human resource issues resolved, financial plans drawn up, staffing decided, assets allocated, rules and regulations set forth, and agreements between the companies signed.

First and foremost, new company territories had to be defined. The three islands of Hokkaido, Shikoku, and Kyushu formed their own natural boundaries, but it was more difficult to divide Honshu Island for the three companies that would operate there. Lines needed to be drawn in areas to which few passengers traveled through. Atami and Maibara on the Tokaido Line and Naoetsu on the Shin-etsu Line were natural choices and were selected immediately.

Although the Nagoya Railway Administration Bureau had managed Maibara Station during the JNR era, privatization placed it under JR West's jurisdiction. Although closer to Nagoya, Maibara sees a greater amount of traffic to and from the western district. Nagano Prefecture, on the other hand, ended up being divided among all three companies, prompting opposition from the local communities.

Next, we had to deal with the *Shinkansen*. The General *Shinkansen* Operation Control Center was located in Tokyo, with large rolling stock depots in Tokyo, Osaka, and Hakata. If the border of the conventional lines were used as a reference, JR Central would be left with neither an Operation Control Center nor a rolling stock depot. This presented a problem.

If JR Kyushu held onto the Hakata depot, JR West would be left without the ability to even develop *Shinkansen* operation plans. Ultimately, JR Central took control of all of the Tokaido *Shinkansen*, and JR West took all of the Sanyo *Shinkansen*. Even this compromise drew strong protests from Kyushu.

The real issue, however, was where to draw a borderline inside a station with both *Shinkansen* and conventional lines. In Tokyo Station, for example, JR Central property would include platforms and other facilities of the Tokaido *Shinkansen*, as well as the land on which they were built. The remainder of the station would belong to JR East. At JNR, we referred to this method of classifying property as the 'Raindrop Method.' All land on which rain dripping from the Tokaido *Shinkansen* tracks fell became the property of JR Central.

The Tokaido *Shinkansen* tracks 14 and 15 at Tokyo Station were not so easily classified. These tracks were originally built for a Tohoku-Joetsu Line platform. Around 1975, however, the Tokaido *Shinkansen* was constantly running behind schedule, and the decision was made to go ahead with the construction of a platform so that Tokaido Line trains could expand into this platform. Viewed from above, it is obvious that the tracks on the north side of this platform merge into the Tohoku and Joetsu Lines. The plan at the time was to use this platform for trains running through between the Tokaido and Tohoku-Joetsu Lines.

The fact that they were initially intended for the Tohoku and Joetsu Lines but ultimately used for the Tokaido Line could not be denied. In the end, the platform and corresponding tracks became the property of JR Central, while the land on which they were built was given to JR East. This is the only instance where the Raindrop Method was not used.

We sometimes receive comments about operating a through operation service between the Tokaido and Tohoku-Joetsu *Shinkansen* Lines. We ultimately concluded, however, that through operations would be extremely difficult. First, these lines operate on different electrical frequencies. Facing the same problem, Nagano *Shinkansen* engineers developed rolling stock that could operate on either frequency. Through operations between the Tokaido and Tohoku-Joetsu Lines would require similar railcar equipment. Moreover, The Tokaido *Shinkansen* EMUs are not equipped with covers for underfloor equipment and other devices used to deal with the heavy snowfall on the Tohoku and Joetsu *Shinkansen*.

The most difficult issue to work around, however, would be the diverse train configurations. A different number of business-class cars in different positions and different seating arrangements throughout the trains would complicate matters considerably. So many trains run on the Tokaido *Shinkansen* that at certain times there is no room to accommodate even one addition. Through operations would, in reality, mean operating Tokaido *Shinkansen Hikari* and *Kodama* trains on the Tohoku and Joetsu Lines. The different train configurations, however, would make this extremely difficult. Trainsets operation is often changed when snow, earthquakes, or other disasters disrupt train schedules, and blending trains of different configurations into this scenario would further complicate matters. With the majority of passengers embarking and disembarking in Tokyo, the disadvantages of through operations were deemed to outweigh the advantages.

As might have been expected, one platform could not handle all of the Nagano *Shinkansen* passengers when the service was introduced in 1997. Nevertheless, Tokyo Station could not accommodate yet another *Shinkansen* platform.

To solve this problem, the Chuo Line platform was raised a floor higher, and the Keihin Tohoku, Yamanote, Tokaido, and other lines were each bumped to the next platform to make room for the new *Shinkansen* platform. Measures this elaborate just to add a platform would not be necessary in any other country. Construction costs alone totaled 43 billion yen.

9.2 Higher Transport Volumes, Stable Ticket Prices

The privatization of JNR was even more successful than anticipated. JNR that had seen a deficit of five billion yen a day by the end of its reign has been transformed into a group of companies that turn a daily profit of 600 million yen today. The JR group as a whole, however, still maintains a debt of 12.4 trillion yen, twice its total revenue, and pays 800 billion yen in annual interest payments. In contrast to the 600 billion yen in national subsidies that JNR received in 1985, its last year of operation, JR pays 200 billion yen in taxes a year to the Japanese government.[1]

How was such a turnaround possible? Why hadn't JNR been able to accomplish this itself? To analyze these points, we must look at the substance of the JNR reform.

In terms of finances, the reason that JR was able to turn a surplus after so many years of its predecessor operating in the red is pretty clear. After privatization, the number of passengers traveling on JR increased significantly. We saw greater revenue than we had anticipated, debt has been reduced to less than half of the JNR era, and we made significant progress in rationalizing operating systems.

Let's look at this in a bit more detail.

JR East's transport volume increased by 22%, quite a bit higher than estimates at the time of privatization.[2] Like the hotel industry, railways are a large-scale process industry. Increased rates of operation are directly linked to profits, while a decrease in the number of passengers makes it difficult for a company to work itself out of deficit operations.

The time was also extremely favorable for the privatization of JNR. The stable political environment and booming economy ushered in an age of 'Japan as Number One.' The wind was literally at our back, and successful reform may have required not only the enthusiasm of the people involved but also a kind of divine grace.

JR transport volumes increased for the first six years after privatization and have stayed at about the same level since. With the bursting of the so-called bubble economy, most Japanese industries saw considerable declines in sales. The fact that JR has maintained its level of revenue in the midst of the decline indicates a fundamental change, rather than simply a temporary increase in passengers.

Another significant factor is that fares have increased very little in 12 years. Although the railway companies in Hokkaido, Shikoku, and Kyushu have raised fares once in the face of extremely severe business conditions, the three Honshu JR companies have never, with the exception of the introduction of the consumption tax, raised their fares. During the last ten years of operation, JNR raised its prices eight times, increasing fares by 76%. Sound JR management has done away with the need to increase prices, and stable ticket prices have contributed to sound management.

JR companies also provide their passengers with much better service, including train speed and the attractiveness of train car design. There are still many areas in which JR does not meet the standards of other service industries, and it could be said that the service during the JNR era was just too bad. Still, I believe better service is another reason that we are seeing more passengers.

---

[1] In FY2019, the JR Group as a whole was making a profit of about 1.9 billion yen per day. However, JR Hokkaido and others, which have many underpopulated areas, are in a difficult business situation; JR's overall long-term debt has further declined to about 7 trillion yen, and combined with the subsequent decline in interest rates, interest expenses have fallen to about 130 billion yen. Taxes paid to the government amounted to about 300 billion yen.

[2] As of 2021, the traffic volume of JR East was 6,688 million passengers, an increase of 36% since privatization in 1986.

Around the time that JR East began to see a steady increase in its passengers, I asked Masatoshi Ito,[3] President of Ito Yokado and one of the JR East Board of Directors,

How he thought this could be explained.

He answered unhesitatingly,

Because service has improved, and service is the key to business.

His answer seemed to me to capture the essence of the service industry.

Even with all of the other factors, debt reduction was key. No company can stay in business when its debt is seven times greater than its annual revenues. The three Honshu JR companies began operating with 14.5 trillion yen in debt. Annual interest alone came to 950 billion yen.

Later, the companies bought *Shinkansen* ground facilities that were originally leased. These expenditures brought the debt total to 15.6 trillion yen or 4.8 times annual revenues. Despite these bleak figures, we were able to bring the company back into the black. In 1985, JNR faced debts that were seven times its revenue and posted a 1.8 trillion yen deficit.

After privatization, JR East focused all of its efforts on debt reduction. Capital investments were carefully screened, and construction and material costs were cut as much as possible. Material inventories were reduced to half of their JNR size. As a result, we reduced our debts by 1.5 trillion yen over eleven years, which, I believe, is a record-setting for Japan.

If these measures had not been taken, JR East would have seen very little profit. It is the instinct of government-run companies to continue to pursue capital investments even in the face of enormous debt. The top priority for a private company is to strengthen its corporate constitution.

This meant rationalizing the railway system, as well. JNR employed a staff of 420,000 in 1979. Today, however, the seven JR companies employ a total of 190,000 persons, despite the 22% increase in train kilometers.[4]

JR East was formed with 80,000 employees, 72,000 of whom worked in the Railway Department. The company incisively introduced automated ticketing machines and computer systems and strenghtened and improved facilities to further rationalize operating systems. The result is a staff of 55,000 in the Railway Department and an increase of 150% in productivity over the preprivatization level. We are finally closer to the standards of other private railway companies.[5]

---

[3] Masatoshi Ito (1924–2023) was a Japanese entrepreneur. He was the founder of Seven & I Holdings Co., Ltd., which includes the Ito-Yokado supermarket and the Seven-Eleven Japan convenience store chain (annual sales of about 18 trillion yen). He was also a former president of the Japan Chain Stores Association.

[4] As of 2019, the staff numbers for railway operations at JR companies were 109,000, and the number of train kilometers had increased by about 30%.

[5] As of 2022, JR East alone (not including affiliated companies), had approximately 48,000 employees, further increasing its productivity.

9.3   Successful Privatization

There were two major factors behind the success of the JNR reform. First, the privatization and division were so extensive in scale that they turned the common sense held by most of the railway community on its head. Second, the officials in charge of the newly formed JR companies were extremely enthusiastic. The reform was more than organizational restructuring; it was a cultural revolution.

From the first day of their existence, the privately operated JR companies faced the danger of bankruptcy. The JNR reform did entail a certain type of bankruptcy, but it was distinct from the severity of bankruptcy in the private sector. The feeling at JNR was that public service was being performed, which made the deficit inevitable. People figured that, ultimately, the government would somehow take care of it. This atmosphere, of course, resulted in tragedy. The first step toward privatization was taken when we resolved never to face bankruptcy again.

We were not, however, brimming with confidence. There was a distinct tone in the media that nothing could bring a company like JNR back into the black. Those who criticized the JNR reform seemed to be waiting for us to fail. We were intentionally labeled: "Once JNR is broken up, there will be no more trains passing through to another company;" "If JNR is privatized, profits will come first, and safety will suffer;" "All local lines running at a loss will be abandoned".

We realized that the key for JNR reform to succeed in this atmosphere of doubt and criticism was to make sure that a considerable number of people would get off their trains feeling pleased that JR was no longer JNR.

Soon after privatization, JR East executives met to brainstorm which aspects of JNR were thought poorly of. There were various suggestions. Service at train stations was poor; railcars were old; stations were shabby; prices were high; the congestion during rush hour was terrible. One person mentioned the dirty toilets. We decided that this was where we would start. The head of the department in charge himself took his staff to Higashi Nakano Station and spent three hours cleaning the toilets.

He mentioned in his report that, when drinking with his staff that evening, "even the sake smelled bad".

We not only cleaned toilets but also spent money on toilets to improve their reputation. Five billion yen was spent on these improvements and earned us the commendation of the Japan Toilet Association in 1988.

Toilets, however, were not the top priority at JR East. The most important issue we faced was how to revamp and improve service on the railways, our core business.

Toward this end, I distributed the following memo.

JR East Railway Charter.

1. To break out of mere transportation to create new needs as the technical service industry.
2. JR East aims for first-class service, first-class technology, and first-class design.
3. JR East to always remember that safety and reliability are the top priorities of a railway business.

4. Service businesses in the future to have sensitivity to anticipate passengers' needs and aim to design services, spaces, and lifestyles.
5. To keep in mind that service business must bring a hit product to the market each year.
6. The railway business to focus on breaking out of the labor-intensive industry and transforming JR East into a modern system industry before the next century.
7. To develop projects organically centered on the railway business in order to form a comprehensive service industry.
8. Future railway businesses to aim for community-based, people-friendly railway services.
9. To utilize employee creativity and establish a corporate culture that its employees are enthusiastic about and take pride in.
10. Companies of the future have the fundamental principle of corporate culture, beauty, and creativity based on sound management.

Readers will perhaps see this charter as a mere game, and it is true that employees did not necessarily accomplish all of these objectives. The charter does, though, capture the spirit and aspirations at JR East just after privatization.

Our first hit product came in the form of the *Hokutosei*, an overnight sleeper train between Ueno and Sapporo. The Seikan Tunnel was completed exactly one year after JNR was privatized. Taking 14 or 15 years to build, this tunnel represented an enormous investment of 1,039.2 billion yen. However, the environment surrounding the railways at the time when this mammoth project started had completely changed.

Most people travel between Honshu and Hokkaido by air; flights between Tokyo and Sapporo carry more passengers than on any other route in the world.[6] Only a small percentage of travelers use the train for this trip.

Even with the Seikan Tunnel in operation, trains take 16 hours to reach Sapporo from Tokyo. The same distance is covered in one hour and 20 minutes by air, and there is little difference in fare. The media have advocated using this tunnel as a storage base for oil. However, it did not change the fact that the completion of this difficult project and the first train schedule revision for JR East since its formation were done.

9.4 *Hokutosei* **Popularity Pleasantly Surprises**

We wanted to try something and we took a hint from the resurrection of the *Orient Express* in Europe. We decided to introduce an overnight sleeper train that, while not perhaps as extravagant as its counterpart, would give passengers the luxury of leisurely travel. Actually, we were not at all confident that the train would be a success. Rather than investing in the design of new railcars, we chose to recycle old ones. Dining cars that were no longer needed on limited express EMU trains were used, which is why the roof of these cars is lower than all the rest (Fig. 9.1).

---

[6] Please see note 10 of Chap. 1.

Fig. 9.1 *Hokutosei*, the overnight sleeper train that became popular. *Photo* provided by the Railway Museum

But we never know unless we try. When we began selling train tickets one month before the departure date, to our surprise, tickets were immediately sold out. Sales continued at this rate until people began talking about how difficult it was to reserve a seat on the *Hokutosei*. Then it became even more popular.

The computer system to reserve *Hokutosei* tickets for stations around the country is activated at precisely 10 a.m. one month before the day of departure. The 'Royal Rooms,' providing first-class private accommodation for one person, were particularly sought-after, but with only two Royal Rooms on each train, tickets were extremely hard to get. It ended up being a case of first come, first served.

To reserve a seat, a staff in the ticket office would enter various information—the date, train number, type of seat, number of passengers, and point of embarkation and disembarkation—into the station computer terminal. The ticket office staff who began to input the data once the computer system was up and running found that they were already too late; the reservation would have gone to another station.

Instead, they would enter all of the necessary data into the terminal before 10 a.m. and sit waiting with their finger poised above the send button. The moment the clock struck 10 a.m., they would punch this button. I saw this scene played out at many stations. Even with these elaborate preparations, it was difficult to secure a reservation, and ticket office staff actually shouted in excitement when they were able to reserve a seat.

This experience was a pleasant surprise. We learned that passengers are not only attracted by faster speeds and new train designs but also by the buzz created by a certain train.

Subsequent products included the *Super Hitachi, Super View Odoriko, Narita Express,* Yamagata *Shinkansen Tsubasa, View Sazanami,* double-decker *Shinkansen MAX,* and finally, the latest *Komachi* and *Asama.* I'm not sure how many times we tried to replicate the success of the *Hokutosei,* but we rarely failed to elicit a strong reaction from our new trains. These successes only reinforce the idea that a railway is more than just a transport industry. After all, even the railways are a service industry, and if they cannot constantly develop new services for their customers, they cannot continue to exist as a company.

Safety, however, is the most important element on the railway. While JNR did work on the safety of its operations, there are quite a few points that still bother me. For instance, not enough was done to improve the Automatic Train Stop (ATS) devices. The Mikawashima accident in 1962 prompted JNR to install ATS devices on all of its lines, but these devices were entirely inadequate. This system did not automatically stop the train when there was danger; it merely issued an alarm to the driver when the train approached a stop signal by using an onboard warning device.

The Automatic Train Control (ATC) device ensures a high level of safety for *Shinkansen* trains. Without exaggerating, the ATC system forms the safety of *Shinkansen* trains. Large-scale private railway companies have also installed various ATS and ATC devices that function similarly to *Shinkansen* ATC. The mix of freight and other types of trains made the ATC and high-performance ATS systems difficult to use, so JNR stuck with its inadequately functioning ATS system. This was the reason behind the freight train collision accident where a tanker burst into flames and caused a major fire at Shinjuku station in 1967, as well as a number of other accidents. The technology for the new ATS system with its high-level safety features was ready for use at the time that JNR was privatized.

At Ueno Station, two trains just missed being involved in an extremely dangerous collision on July 3, 1988. An incoming *Asama* limited express failed to stop at the stop signal at the entrance to the station and nearly collided with an EMU train that was just departing from Ueno Station.

Soon the Board of Directors resolved in September to install a new ATS-P safety device on the Tokyo Chuo Line and other lines that are subject to extremely heavy traffic. Tragically, however, two EMU trains collided at the Chuo Line's Higashi Nakano Station just three months later on December 5. This tragic incident claimed the life of one person (Fig. 9.2).

We subsequently rushed the installation of this new type ATS-P at these stations, and almost every line near Tokyo is equipped with this safety device today. The capital investment on this project alone totaled 80 billion yen.

Since it was formed, JR East has invested 2.2 trillion yen for safety measures in the broadest sense, including the replacement of old railcars. I have no doubts that this exceeds the amount spent during the JNR era. As a result, the operating accidents

Fig. 9.2 December 5, 1988,
Collision at JR Higashi
Nakano Station brought the
third car smashing into the
second one. *Photo* provided
by Kotsu Shinbunsha

officially reported to the Ministry of Transport have been reduced by more than half,
and level crossing accidents to one-third of their previous numbers.[7]

The bulk of these investments were not made in compliance with government regulations or administrative orders; JR East invested in safety of its own accord. Transportation companies' biggest fear is a serious accident. Privatization of a company
means standing on the principle of self-responsibility, and safety is its starting point.

In addition to improving safety, we were adamant at the time of the JNR reform
that we would never fall back into our previous working conditions. This resolve and
regrets from the past were strongly held by JR management and, I believe, by the
labor unions and most of the employees as well.

One particular memory from about seven or eight years before the JNR reform
has left an indelible impression on me. My daughter, a junior high school student at
the time, came home one day looking very pale. She looked up at me and screamed,

Dad, why do you have to work for JNR?

---

[7] As of 2021, "Operating accidents" had been reduced by one-third and level crossing accidents by
one-fifth, respectively.

At that time, JNR had just raised the fare level and in addition to that, train operations were being disrupted daily by a continuous 'work-to-rule' sabotage. My daughter's classmates were apparently bullying her at school.

I had no idea what to say to her, and I'm sure quite a few other JNR employees were in the same position. I had a feeling of being shrugged off by the rest of the world.

In one sense, JNR taught JR companies a valuable lesson. In working to establish JR, we resolved to dismantle JNR traditions and practices, and to learn from the principles of private business.

We would end the archaic system of an elite class of employees appointed to positions of intense responsibility at extremely young ages and shuffled from position to position every year or two. We would prohibit the 'descent from heaven' system under which retired executives went to work for private companies, which was natural considering that we were no longer a part of that 'heaven.' We would reverse the JNR practice of rarely employing women, and instead actively seek them out. We would also, of course, introduce promotion exams and performance assessments. In the era of JNR, strong opposition from labor unions had drastically curtailed the use of these tools.

We would also weed out old business practices. Bargaining between acceptance of descent from heaven and business order was now forbidden. The custom of distributing orders for services and materials equally among certain preselected companies would be eliminated. If the quality was high and the price low, we would purchase products continually from the same company, from new companies, or from companies overseas.

The JNR practice of purchasing exclusively from domestic companies has been dismantled. Door opening/closing devices for the doors used on EMUs on the Keihin Tohoku Line are manufactured in France, as are the seats and windowpanes on the Narita Express. Parts for the new *Shinkansen* EMU train electronic control device and brake mechanism are produced in Germany. These are only a few examples. This is a big difference from the days of JNR when all domestic products were the rule. Although materials purchased from foreign companies totaled only 200 million yen in JR East's first year, this sum now reaches nearly 10 billion yen.

Reduction of construction costs also featured significantly. The 'bubble economy' had burst, and construction costs for buildings and other structures had fallen by 40 to 50%. The construction cost of the Akita mini-*Shinkansen* came in at 14% less than planned. There are numerous examples of JNR construction costs exceeding estimates, but I do not remember a single occasion when construction cost considerably less than the budget. The essence of the JNR reform dealt with putting an end to the obsession with acquiring and using up budgets, sectionalism, the 'descent from heaven' system, and obduracy with precedents.

9.5 Japan's Privatization Impacts the Railway Community

The privatization of JNR had a considerable impact on railways throughout the world. After building the world's first high-speed railway, *Shinkansen*, the world's largest transportation company, collapsed, leaving behind record debts, only to successfully rebound. This story naturally attracted the world's attention.

The year after the JNR reform, I visited Paris again for the first time in 10 years. In September 1988, a bon voyage ceremony was held to commemorate the unbelievable trip of the *Orient Express* from Paris to Tokyo. Before the JNR breakup, Fuji Television had consulted me about this event. My research showed that the trip was technically possible, and it was realized.

The train was to depart from Paris, travel through Germany and Poland, enter the Soviet Union, and make its way across Siberia and through China to arrive in Hong Kong. Passenger cars would then board a ship carrying them to the Port of Kudamatsu in Yamaguchi Prefecture, where the *Orient Express* started its travel on Japanese rails and finally arrived at Tokyo Station. The Paris-Hong Kong leg would cover 15,000 km and take 20 days. The gauge would change twice along the way, requiring two bogie changes. Since Japanese railcars are nearly the same size as European cars, the meter gauge tracks would not present a problem as long as narrow gauge bogies were brought along (Fig. 9.3).

Thirty-six passengers boarded this train between Paris and Hong Kong. The cost of a single ticket was 2.5 million yen. Once in Japan, the itinerary included a variety

Fig. 9.3 The *Orient Express*, a dream train, running with Mt. Fuji in the background. *Photo* provided by Masatoki Minami

of tours that would take the train from Hokkaido to Kyushu. Most of the excursions proved extremely popular and were fully booked when people realized that the *Orient Express* dream train would be traveling through Japan.

In Paris on the morning of September 7, the Lyon Station departure board read 'Orient Express to Tokyo,' marking the departure of what was undoubtedly to be a one-time-only train ride. At the platform sat a steam locomotive heading up the electric locomotive and 11 dark blue cars making up the *Orient Express.* SNCF had prepared this steam locomotive, especially for this commemorative train event.

I was also on board the train between Paris and its first stop at Rheims. It was indeed like a dream come true.

That evening, I received a dinner invitation from Jean Philippe Bernard, the International Bureau Chief at SNCF. He was much calmer and warmer than typical elite Frenchmen. When we turned from the *Orient Express* to the topic of the *Shinkansen* and TGV, he said, The Japanese railway has made two extremely significant contributions to the world's railways—the *Shinkansen* and the JNR privatization. The success of the *Shinkansen* saved passenger trains all over the world from extinction. The JNR privatization was, I'm sure, a difficult experiment, and SNCF has no intention at the moment of taking the same path. But, at the very least, privatization has shown us that it is possible for railways to turn a profit (Fig. 9.4).

Fig. 9.4 Mr. Albert Glatt, President of Intrafrug. A.G. (left), Mr. Shuichiro Yamanouchi, Vice president of JR East (center), Mr. Isamu Yamashita, Chairman of JR East (right) at Keio Plaza Hotel. *Photo* provided by Masatoki Minami

The JNR reform launched a wave of privatization among the world's railways and was an impetus, particularly for those involved in railway policy. In November 1991, the U.S. Department of Transport sponsored an international conference on railway privatization in Washington D.C. Railway representatives from 60 countries were in attendance. One of the keynote speakers, UIC Secretary General Walrave gave a particularly hard-hitting speech in which he expressed the feelings of French railway people exasperated by the trend toward the privatization of railways.

> Mr. Yamanouchi from Japan has joined us at this conference today. Let me say that the JNR privatization is a fake. The government still owns all company stocks, and the debt problem has not yet been resolved. These conditions cannot be called privatization.[8]

The recent move toward privatization, however, cannot be denied. In June 1991, the European Union (EU) announced its basic policy regarding the management of Europe's railways. In its famous Directive 440/91, the EU stipulates that: (1) railway companies in each country will promptly resolve their debt problems and put their finances in order; (2) governments will intervene less in operations and will guarantee independent railway management; (3) railway companies will be divided into companies managing tracks and other facilities and companies operating trains; and (4) railway lines in each company will be open to requests from all railway companies, including those in other countries.

9.6   Vertical Separation in Europe

In contrast to the regional basis for how JNR was divided when it became JR, Europe divided its railways into companies in charge of managing and repairing ground facilities and companies responsible for operating the trains that would run on these facilities. If the Japanese method is referred to as regional division, the European method is one of vertical separation.

Transportation policy became an extremely important issue as Europe pursued integration. The 1957 Rome Agreement, which laid the foundation for European integration, set forth a collective transportation policy for this region. Trains, as well as airplanes and automobiles, had been structured at the national level and needed to be adapted to the European unit as a whole. By vertically dividing railway management, decision-makers felt, all companies, regardless of national origin, would be free to participate in the railway business, and this competition would serve to enhance service and increase productivity.

The concept behind this was that in Europe, where transportation density is as low as in Hokkaido, it would be nearly impossible for a company that owned every aspect of the railway down to its infrastructure facilities to turn a profit. For this reason, governments often end up supplying the money for infrastructure maintenance and

---

[8] JR East, JR Central, JR West, and JR Kyushu are already publicly listed, and the government no longer owns their shares. In addition, the long-term debt of 25.5 trillion yen, which the government inherited, has been reduced to approximately 17.8 trillion yen.

construction, even after the railway has been vertically separated and privatized. The same can be said for airports and highways.

Losses on local lines are another problem common to all countries. More and more railway companies in Europe are receiving necessary subsidies from local governments to operate local railway lines. These companies include, but are not limited to, railway companies once operated by the government.

Sweden was the first to privatize and vertically separate its railways. The EU Directive was modeled on the Swedish example. In 1988, the Swedish State Railways (SJ AB) was divided into a public entity that would maintain railway infrastructure and a private company that would operate the trains. All infrastructure expenses currently fall to the Swedish government.

In 1994, DB was privatized, and a joint-stock company was formed. The railway was not divided up at first. In 1999, it was separated into a number of companies responsible for long-distance passenger travel, local passenger travel, freight transportation, and infrastructure. The German government took over the entire four trillion yen debt left by DB. Private railway companies now negotiate with local governments about subsidies to assist with the operation of local trains. Germany may offer the best example of a truly European style of privatization.

The most dramatic national railway reform, though, was implemented by the UK. Railway infrastructure was to be retained by a company called Railtrack, while different companies would be in charge of tracks, and signal equipment maintenance. A separate company was established to lease rolling stock, while others would be responsible for repairing them. Train operations would be divided by track section and 25 railway companies would be established. By 1997, the privatization of British Rail was nearly complete, at which time a hundred new companies had been established and the stock listing completed.

The operation of British trains fell not only to Virgin Airlines or British bus companies but also a French waterworks company and a U.S. freight company. Moreover, British Airways, SNCF, the Belgian National Railway (NMBS/SNCB), and other companies, as well as National Express, invested in the London-Continental Railway Company, which went on to construct the high-speed railway between London and the Channel Tunnel.

Opinion is currently divided about this radical privatization and division. While some point to the fact that railway services have improved and trains more often run on time, others maintain that the government's financial burden has increased since privatization. Critics voice misgivings about whether the investments necessary for the long-term will be made. The general sense in the railway community, however, is that these newly formed companies are off to a better start than many feared.

Following the examples of Sweden, Germany, and UK, most European railways have taken steps toward privatization and vertical separation, although the methods they are using vary. This includes Italy, Holland, Switzerland, and even former Eastern European countries that are moving toward privatization.

The country most strongly opposed to the privatization trend is France. No steps have been taken to privatize SNCF. With the country's culture of strong government regulation and bureaucratic authority, railway executives seem to feel that privatization would reflect a loss of standing. Labor unions have also fiercely opposed privatization from the outset. In 1997, a demonstration protesting EU railway policy was held in Brussels, the capital of Belgium and the location of the EU Headquarters. Other countries offered little support for the protest, which only served to reinforce France's feeling of isolation.

That same year, SNCF finally roused itself to implement vertical separation, but not privatization. Although it retained control of the railway's Operations Division, the government-run French Rail Network (RFF) was formed to take possession of infrastructure, which they would, in turn, rent to SNCF. The government took over approximately three trillion yen in SNCF loans, totaling three-fourths of its debt. This condition was needed to convince reluctant labor unions and avoid violating the EU Directive.

Dissenters, however, maintain that this type of reform is merely a deception. They point to the fact that the fees charged by the RFF for renting facilities to SNCF are inordinately low. They call this an attempt to continue the government subsidies with a better appearance. Critics also maintain that RFF itself is a sham and will commission SNCF to handle facility repairs, inspections, improvements, and all other work. France continues to oppose opening its railway lines to foreign railway companies, which could be said that this is a sham reform.[9]

The wave of privatization is not limited to Europe. Countries in Asia and Oceania, as well as South America, are rushing to privatize their railways. In Malaysia, New Zealand, Argentina, and other countries, privatization is already complete; in some countries, the governments have divided their railways and sold stock into several companies. In many cases, U.S. capital was involved and railway operations were rationalized to arrive at autonomous management.

Even the U.S. has begun reforming its railways. Private companies have always operated railways in the U.S., so privatization is not necessary. Large-scale government deregulation, though, has revitalized these companies. Transport volumes have increased 30% or more, and they now enjoy significant profit margins. CSX, a freight railway company operating lines along the East Coast of the U.S., generates only half the revenue of JR East, but sees nearly 50% greater profits. The major U.S. railway companies all specialize in freight transport, while passenger transport is dependent on government subsidies.

More than anything, deregulation liberated the railways from fixed pricing, so that companies sending a specified number of containers are now able to contract at low fees. U.S. freight companies also apparently maintain long-term contracts

---

[9] The French National Railways (SNCF) operated the railway for 18 years based on the 1997 vertical separation reform of the national railway system. However, a re-reform took place in 2015 due to conflicts with SNCF, because the French Rail Network (RFF) was set up as an independent organization. This re-reform resulted in an operating structure in which SNCF Mobilité and RFF are subsidiaries under the umbrella of a holding company. Furthermore, in 2020, under SNCF (the parent company), which is wholly owned by the state, the company were reorganized into five subsidiaries by business field: Re'scau including Gares & Connexion, Voyageurs, Keolis, Rail Logistics Europe, and Geodis.

covering three- and ten-year periods. There is also progress being made in terms of rationalizing operations. In the past, one train required rotating shifts for 56 crews to travel from Chicago to Los Angeles. Today, just 22 crews do the same amount of work.

U.S. freight trains also operate larger units than Japanese railways. While freight trains in Japan are only 600 m long, the heavy cargo trains in the U.S., dubbed 'mile trains,' are three times this length. Containers are stacked on top of the base containers so that, with two levels of containers, one train is able to transport 180 containers.

The small size of tunnels and bridges in Japan prohibits freight trains from carrying even one level of international containers.[10] A Japanese freight train can accommodate only 130 compact railway containers.

U.S. railway companies are also merging at a rapid pace. Over 100 major railway companies have now merged into only three or four. The result has been further rationalization by abandoning operations on parallel routes, consolidating switching yards and otherwise eliminating waste. Although U.S. railway companies are now in excellent condition, the recent merger between Union Pacific and Southern Pacific Railways has apparently not been going well.[11]

Russia, China, and India have not yet privatized their railways. What is the situation in these vast countries? The total track length in Japan is 28,000 km. In comparison, Russia's railway totals 87,000 km, while India maintains 63,000 km. While 25 times the size of Japan, China has only about 60,000 km of railway track, and is now laying new lines at a pace of 1,000 km a year.[12]

On the one hand, China is working on construction plans for a high-speed train corresponding to the *Shinkansen* to run on the approximately 1,600 km section between Beijing and Shanghai. At the same time, however, they plan to reduce national railway staff by 1.1 million dramatically over a three year period beginning in 1998. Currently, the 3.5 million railway workers are working on lines three times the length of JR's. This socialist country is undertaking rationalization of a scope far beyond the JNR reform.

9.7 275 km/h: Pride in Privatization

JNR reform also offered a new opportunity to renovate our technology. I can honestly say that almost no technical progress was made during the last ten years of JNR.

It was during this period that most Japanese industries made their greatest progress in technological innovations, which meant that we paid a particularly heavy price for

---

[10] Currently, a container freight car of JR Freight can carry two international containers with a total weight of 20 tons.

[11] The merger of the two companies was completed in 1996.

[12] China is expanding rapidly, with 146,000 km of high-speed rail and urban railway expansion. The number of employees of China's national railway company reached about 1.9 million by 2020.

JNR's inertia. JNR had the money; a capital investment of one trillion yen was made annually. Plagued by financial and labor union problems, management was apathetic toward technological innovation. The loss of inaction in JNR era is significant. JNR technicians were more enthusiastic about acquiring budgets and securing projects than developing new technologies. Tense labor-management relations forced administrators to seek out labor unions and kowtow before them just to conduct a single test run. Each of these factors contributed to the erosion of JNR technology.

Isamu Yamashita, JR East's first Chairman, has described the lamentable conditions at that time. "Looking in as an outsider, JNR technology seemed suspicious. But I had no idea just how bad it was until I went to work for JNR".

The *Shinkansen* was no exception. True, the tracks and electrical equipment of Tohoku and Joetsu *Shinkansen* were sturdier and superior to Tokaido *Shinkansen* line. These improvements, however, were also extremely costly, and I doubt they can be called technological progress. A new large-scale computer system was introduced, but there were few personal computers or even a few fax machines at the front of the field. The abacus was still used to calculate daily ticket sales.

Shinkansen EMUs had not changed much in over 20 years, either. Some changes were made to the exterior design of the 100 series *Shinkansen*, which debuted in 1985, and double-decker business-class cars were added. Only minor technical changes were made, though, to the actual structure. Neither had the maximum speed increased beyond the 210 km/h at which *Shinkansen* service was introduced.

As the deadline for privatization approached, signs of change in the *Shinkansen* system became evident. Around this time, JNR management began to realize that the situation could not continue and reforms would indeed be necessary.

In March 1985, construction on the Omiya-Ueno section of the Tohoku *Shinkansen* was completed, and JNR took this opportunity to increase the speed of Tohoku *Shinkansen* *Yamabiko* to 240 km/h. This was the first increase in speed in the 21 years since Tokaido *Shinkansen* line service was introduced. At the same time, certain *Yamabiko* that had previously stopped at every station between Sendai and Morioka began running non-stop on this section. At last, the *Shinkansen* was living up to its name as a high-speed train.

The speed of the Tokaido *Shinkansen* was increased by 10 km/h to 220 km/h the following year. At that point, the highest speed at which the ATC device could function was 225 km/h, just five kilometers per hour higher than the speed the trains were running at. In the past, the maximum train speed and the speed at which the ATC activated the brakes were both 210 km/h. Trains were difficult to operate at the increased speed because the brakes would be activated as soon as the train reached 210 km/h.

In many cases, trains were forced to run a few kilometers under 210 km/h; 200 km/h was used as the maximum speed to calculate trip time when planning train timetables. This complication prompted JNR to increase the ATC operating speed by five kilometers an hour to permit trains to operate at their full speed of 220 km/h and reduce trip times for train timetables. This posed no safety problems. As a result, 12 minutes were shaved off the time it took to travel between Tokyo and Shin-Osaka, and we had finally accomplished the long-held goal of cutting this trip to within the two-hour range. By thinking slightly outside of the box, our trains were able to reach higher speeds.

It was here that we welcomed the privatization of JNR. We were eager to enter the technology race again and reclaim the top position in the world. Speed is not, as I've said, everything, but we were certainly not thrilled about falling behind France's TGV system. Unfortunately, JNR era inertia had left us with little in the way of new technology or test data. The Tohoku-Joetsu *Shinkansen* of JR East had only recently been opened, and we were not yet ready to develop a new series of EMUs.

We decided instead to research the possibility of overtaking the TGV 270 km/ h with slight improvements to the EMUs that were already operating. Although stability and pantographs presented very few problems during our series of high-speed test runs, the EMUs designed for maximum speeds of 210 km/h were not powerful enough to reach higher speeds.

The cars refused to pick up speed; when they did, ATC modifications became necessary and noise problems were a concern. It was therefore decided that work on increasing overall speed would be postponed. We would focus first on achieving higher speeds on one particular section of the track. This would serve as a first step toward future high-speed operations and provide data in many areas.

The Jomokogen-Urasa section of the track was chosen for this experiment. This section passes through the Daishimizu Tunnel. This long tunnel spans over 20 km, and down trains passing through this tunnel from Tokyo to Niigata must approach a lengthy, downhill grade. Trains would use this gradient to increase their speed.

There was one complication, however. Closer to Tokyo, trains would encounter the Nakayama Tunnel, and in one part of this tunnel, speed would have to be limited to 160 km/h.

The original plan had called for a straight section of track, but a large flood had shocked construction workers during the excavation phase. According to personnel in charge of the site, the mountain contained a huge mass of water. A damaged wall had caused the flood, and workers could do nothing to stop the water. The tunnel had to be re-routed and re-excavated, resulting in a fairly sharp curve that forced trains to pass at lower speeds.

We wondered whether, after slowing down to 160 km/h in this tunnel, we would be able to match the speed of the TGV. Test-running indicated that trains could reach 275 km/h on the descent.

Although it was a mere five kilometers an hour, it was faster than the TGV. Since March 1990, two Tokyo-Niigata *Asahi* trains a day began running on this section at a speed of 275 km/h. Trains traveling in the opposite direction, however, were still unable to increase their speed. Still, until 1997 when the Sanyo *Shinkansen Nozomi* achieved 300 km/h, these were the fastest *Shinkansen* trains (Fig. 9.5).[13]

France, however, had outdone us. Six months before the 275 km/h *Asahi* trains were introduced, trains began running at 300 km/h on the TGV Atlantic Line. JR East's dream of operating the fastest train in the world burst like a bubble.

---

[13] Since 2017, JR East's E5 and E6 trains have been operating at a maximum speed of 320 km/h.

Fig. 9.5   300 km/h *Nozomi Shinkansen. Photo* provided by the Railway Museum

9.8   Elegy to *E-Den*

Once JNR was privatized, names were needed for the newly formed companies. The official names—Hokkaido Railway Company, East Japan Railway Company, and so on—would have to be shortened to make it easier for people in other countries. For instance, the Denden telephone company had undergone privatization earlier, and the nickname NTT had become part of the popular vernacular. Japanese National Railways had been abbreviated to JNR, which was the only name by which it was known overseas.

Of the two candidates, NR for National Railways and JR for Japan Railways, JR was chosen by an overwhelming majority. Each JR company was also to have a distinctive color to symbolize the region in which it operated. Green was selected for JR East to evoke the large natural areas in the Tohoku region; orange for JR Central because of the mandarin oranges grown there; blue for JR West and JR Shikoku in connection with the ocean; and red for JR Kyushu, since it is known as the fire island in Japan.

The color for JR Hokkaido, however, sparked some debate. White, to symbolize snow, was the first color to be suggested. This, however, was strongly resisted by the people of Hokkaido. They insisted that while people in Honshu immediately thought of snow in connection with Hokkaido, in their own eyes, bright green was a more appropriate color. In deference, bright green was selected for JR Hokkaido. On the day following the privatization of JNR, every train was adorned with the large letters 'JR,' which were painted during the previous night.

Soon after, we had to decide what to do with the name *kokuden*, literally "national railways' EMU train." This nickname was no longer appropriate. The problem was that no one had actually chosen the name; it had naturally come into being.

Before *Kokuden*, the same trains had been called *Shosendensha*. This name, literally 'Ministry of Transport EMU trains,' had also just appeared at some point to distinguish between government-operated trains and trains operated by private companies such as Tobu or Odakyu. It was named after the Ministry of Transport, which directly operated the railways. Soon after JNR was formed as a public corporation in 1950 when the Railway Division of the Ministry of Transport became an independent entity and *Kokuden* came into use to refer to the JNR trains. Even so, *Shosendensha* did not disappear entirely, particularly among the elderly, for quite some time. *Shosendensha* referred to all the EMU trains operating on JNR tracks. With the exception of certain lines purchased from privately operated railways, EMU trains ran only on large urban transportation lines in Tokyo and Kyoto-Osaka-Kobe. No EMU trains ran on the Tokaido and Tohoku Lines; the trains on these lines consisted of coaches hauled by locomotives.

After JNR was established, however, the situation began to change. The Shonan EMU train, which debuted in 1956, was entirely different from the traditional *Shosendensha*. Not only did this train travel over a much longer section of track, but it was also equipped with coach-style facilities. Entrance/exit decks were separated from passenger seating areas, and they were equipped with toilets and other amenities. The EMU train came into the world of passenger coaches. This is quite different from the image of a traditional *Shosendensha*.

When the nickname *Kokuden* replaced *Shosendensha*, the term took on a different meaning. It no longer referred to all JNR EMU trains but was limited to those EMU trains used mainly by commuters and other passengers traveling relatively short distances. In these cars, there is no distinction between boarding areas and passenger rooms, four doors open directly onto seating areas equipped with a long bench-style seat lining each side.

JNR employees sometimes referred to *Kokuden* as *geta* EMU trains. This nickname was derived from the casual feeling of these trains, similar to the feeling one has when wearing *geta*[14] rather than more formal shoes.

EMU trains were further developed to include express and limited express trains. Around this time, JNR EMU trains were divided into five broad categories. In addition to the *Kokuden* for short-distances, there were also *Chuden* traveling mid-range distances. The structure is similar to the *Kokuden* in that their doors also opened directly onto seating areas. *Chuden*, however, are equipped with both long bench-style seats lining the sides and smaller two-seater benches placed perpendicular to the sides, facing each other. These cars have three doors, and certain EMU trains of this type are equipped with toilets, as well. The EMU trains running on the Yokosuka and Tohoku Lines are an example of this second category. *Chuden* travel slightly longer distances and so are equipped with facilities midway between *Kokuden* and traditional passenger coaches. It was a mid-term EMU in that sense.

---

[14] Traditional Japanese wooden footgear.

The third category was the express type. These EMUs have coach-style structure, with doors separated from passenger rooms, rows of cross seats, washrooms, and toilets. To improve passenger comfort, these cars were developed with air springs from the beginning. Recently, however, this type of EMU is on the verge of disappearing as express trains themselves have almost disappeared and the in-car facilities are inadequate.

The limited express type is the one in which EMUs are a higher class than the passenger coaches. This type of EMU train debuted on the Tokaido Line with the *Kodama*. This was the first train with air-conditioning in every car and sealed windows that could not be opened or closed. This train significantly altered the Japanese culture of station lunch boxes. More comfortable than on express trains, limited express seats are designed to allow passengers to sit in either direction. The last category was the *Shinkansen*.

After JR companies were formed, we struggled to find a new term for *Kokuden*. Since we were a privately operated company, this term now sounded strange, and we hoped to break loose from JNR's reputation as quickly as possible. A campaign was begun asking passengers for suggestions. We received as many as 60,000 replies, but to be honest, no particular suggestion jumped out at us.

Minden (private EMU train) was the overwhelming leader in the acknowledgment of JNR privatization. Although temporarily appropriate, we doubted the name would stick with people over the long run. Besides, all private railway company trains are technically *Minden*.

The next *Shutoden*,[15] seemed appropriate in the sense that JR East trains operate mainly in Tokyo, but this word is slightly hard to pronounce in Japanese. Certain screening committee members were also concerned that pronounciation on *Shuto* is alarmingly close to *Suto*,[16] and it was rejected. Another suggestion, *Totetsu*,[17] was rejected on the grounds that this was the abbreviation used for the Tokyo Railway Administration Bureau during JNR era, and it sounded too formal.

We began to think about the elements that make a name appropriate. It should evoke the image of an EMU train from JR East in the public's mind. It should set our EMU trains apart from those run by JR Central and JR West. In this sense, *Touden*[18] was the most suitable, but it was already the abbreviated name of Tokyo Electric Power Company. Moving much farther down the list to the 20th most popular suggestion, *E-Den*[19] was ultimately chosen. We were concerned about the pairing of the Roman alphabet with Chinese characters, but then again, Roman letters are often used in Japan. We hoped that this name would help us escape from the prevalent image of JNR as an inflexible, formal company (Fig. 9.6).

Our choice, however, was an utter failure. We were strongly criticized for corrupting the Japanese language, and even worse, the nickname never caught on.

---

[15] It means capital in English.

[16] It means a labor union strike in Japanese.

[17] It means East railway in English.

[18] It means East EMU trains in English.

[19] "E" is from JR East and "Den" is an abbreviation of EMU train in English.

Fig. 9.6 '*E-Den* Platform' sign at JR Shibuya station (The third line from the bottom). January 1990. *Photo* provided by the Railway Museum

Instead, our trains came to be referred to simply as JR. For passengers, the distinction between *Kokuden* and *Chuden* is not important. Today, when nearly all trains operating in Japan are EMU trains, the word *densha*[20] itself is no longer necessary. By May 1996, coaches hauled by locomotives had all but disappeared from JR East lines; the only exceptions being night trains and extra trains.

Neither are passengers concerned with the distinction between *Kokuden* and *Chuden* trains, which has nearly lost its meaning on today's railway system. In the past, *Kokuden* carried passengers traveling relatively short distances, which meant extreme congestion during rush hour. Their long bench-style seats and greater number of doors were designed to increase passenger capacity as much as possible. Doors that slide on the both sides were installed to allow passengers to get on and off of trains as quickly as possible, thereby reducing the length of station stopping time and bringing succeeding trains into the station as quickly as possible.

The concept behind the *Chuden*, however, was slightly different. These trains traveled over slightly longer distances and were designed to ensure as many passengers as possible had a place to sit. Rush hour crowds also had to be taken into consideration, though. By reducing the number of door by one to three doors, long bench seats could be added in the congested areas. This was the original *Chuden* design.

As people began moving out of the heart of Tokyo and residential areas shifted to the suburbs, *Chuden* on the Tokaido, Tohoku and other lines became more crowded. The conventional EMU design did not allow passengers to board quickly enough, and

---

[20] It means EMU train in English.

trains began to fall behind schedule. Although not all JR East officers approved, more long bench-style seats were added gradually. The cross seats originally characteristic of the *Chuden* are only left partially in three cars of 15car trainsets of the newest 217 series EMU in operation on the Yokosuka and Sobu Lines.

Kokuden and *Chuden* also differed in terms of car body shape. Car bodies on *Chuden* were slightly wider, which made the lower parts of these cars curved. This modification allowed wider aisles to be secured between the cross seats extending out from the sides. With more and more long bench-style seats on *Chuden*, the car body shape no longer required modification. Wider cars would also significantly increase passenger capacity on *Kokuden* EMU trainsets. *Kokuden* with the curved car body shape of *Chuden* were introduced. As a result, capacity on these 10car trainsets was increased by 80 passengers, which helped alleviate the problems with rush hour crowds.

Although the modifications were necessary, it seemed sorry for passengers to be traveling these longer distances with only long bench-style seats to sit on. We increased the number of *Commuter Liners*, EMU trains on which passengers could sit while they commuted to work, and developed special double-decker EMUs. Rapid service trains using these special EMUs are operated during afternoon hours with fewer passengers.

The EMUs introduced after transformed to JR with the most significant impact, however, was undoubtedly the six-door no-seat car. The name is somewhat of a misnomer, though, because these cars are equipped with electric folding seats. The seats remain folded away between the time the first train departs early in the morning and 10 a.m. when the rush hour is over. These cars were adopted in an attempt to alleviate the extreme overcrowding on the Yamanote Line during rush hour. The number of doors on these cars was also increased to reduce the delays caused by the amount of time it took passengers to get on and off these trains. JR East was severely criticized, however, for instituting 'cattle cars' and 'not treating our passengers as human beings'.

We were, however, left with no other option. Since its formation, JR East engineers had thoroughly studied the possibility of adding another car to the 10-car trains on the Yamanote Line. They had somehow managed to do this, and this six-door, no-seat railcar was added as the eleventh car on the train. The total number of seats on the trains as a whole, therefore, remained the same (Fig. 9.7).

If the public accepted this new type of car and allowed us to replace all of the cars on these trainsets, it would be possible to reduce the length of station stops and increase the number of trains operating during rush hours. The initial reaction of the media was merciless, and most initial passenger comments were also critical. Comments in favor of these cars, however, gradually increased, and some passengers even suggested using these railcars on other lines as well. The six-door, no-seat railcar has now been added to trains running on the Keihin Tohoku Line.

Challenging the taboos of JNR era was another important issue for us after JNR reform. Six-door cars, rapid service trains on the Keihin Tohoku Line, Saturday rapid service trains on the Chuo Line, automated ticket machines and non-stop *Shinkansen* service: all of these had been off-limits at JNR. New products and services will, of

Fig. 9.7 One six-door, no-seat railcar was added to Yamanote line trains. The photograph shows the car with its seats folded. *Photo* provided by Kotsu Shinbunsha

course, never please everyone. Some people will welcome, and others oppose, rapid service trains that pass through certain stations without stopping, for example. The attitude at JNR had been that it was safer to do nothing at all.

9.9 New Train Contributes to the Environment

Although not exactly breaking through taboos, the proposal to develop a train with half the lifespan and half the weight at half the cost of conventional EMU was a radical departure from the rigid mindset of the past. At JNR, trains were to be used as long as possible. In order to convince executives to approve the replacement of an old train, it was typical for workers in the field to bring photographs of dilapidated car bodies to upper management.

I thought this was a bit odd.

Decades ago, of course, people would repair household appliances and patch up old shoes for as long as they were still useful. Back then, repair costs were extremely inexpensive. Today, though, repairs are very costly, and major repairs are almost unheard of. On the railways, however, all major rolling stock parts are dismantled every six years for inspection and repair. So we began to doubt that ours was in fact the proper procedure.

In the 'real' world, no one regularly dismantles his or her car or television for inspection. Household appliances and EMUs are, of course, completely different animals. Anytime, though, we find ourselves using practices left over from a long-past era when products were cheap and repair expenses even cheaper, something we need to reevaluate.

In researching the cost of JR East ground facilities and rolling stock inspections and repairs, we found that the money spent in this sector, including personnel expenses, came to one-third of our total operating expenses. Not only is it a matter of money; but keeping old rolling stock in use over long periods of time also means poor passenger service and stalled technological progress.

In many cases, JNR EMUs were operated for about 25 years before being scrapped. Older railcars were often removed from urban lines and relocated to local lines in rural areas. Soon after JR East was formed, we resolved to introduce new EMUs on the Nanbu Line, and the people living in this area laughingly thanked us for the first new rolling stock in 50 years. As a first step, we hoped to cut the lifespan of our EMUs by half. We knew, though, that others at JR would consider this suggestion crazy.

We realized, though, that the Investment Management Department would not complain if we proposed at the same time to cut procurement costs in half as well. With overall expenditures remaining the same and repair expenses decreasing, they would be more profitable. Although reducing EMU costs by as much as half may sound ridiculous, many people, including those who had retired from JNR, have pointed out that construction and materials prices were considerably higher at JNR than at private companies operating in intensely competitive environments. Radical EMU design changes and large-volume orders based on a dramatically lower lifespan would also help make this possible.

To these two modifications, we added the concept of cutting train weight by half. This would result in lower expenses for materials, as well as power cost and railway track repair costs.

The 209 series EMUs, introduced on the Keihin Tohoku Line in 1993, were the first railcars to incorporate this new concept. Although we were not successful in halving either the cost or the weight, both factors were reduced by about 30%. With this accomplishment, we began new research to prepare for the next step (Fig. 9.8).

The new development concept was criticized as going against the grain in an era that placed importance on protecting the environment. Our new EMU cars, however, actually contributed to environmental protection. Their lighter weight and innovative electric power control system noticeably reduced the amount of electricity required to operate them. Replacing an old 10-car trainset 103 series alone conserved the amount of electricity normally consumed by 800 households.

Recycling will compensate for the shorter railcar lifespan. On a weight basis, we are already recycling 90% of scrapped materials from these EMU cars. In terms of improving the quality of service, lowering costs, and reducing environmental impact, this new concept offers significant technological innovations.

For JR East, enhancing urban transportation is a more critical issue than the *Shinkansen*. Of all JR East passengers, 75%, or 13 million, use our trains to travel

Fig. 9.8 This 209 series train debuted on the Keihin Tohoku line. *Photo* provided by the Railway Museum

relatively short distances of 50 km or less, to commute to work or school, for example. This is twice the total number of passengers on French and German railways. Generating approximately 650 billion yen in revenue, more than that generated by the *Shinkansen* trains,[21] short-distance urban transportation is crucial to our company. JR East is essentially in the urban transportation business.

Over the 11 years since it was formed, JR East has invested 2.2 trillion yen to increase transportation capacity, enhance facilities, implement safety measures, and take other actions on its Tokyo area lines. Specifically, the Saikyo Line was extended to Shinjuku and Ebisu, and the Tohoku-Takasaki Line to Ikebukuro; additional Commuter Liners are in operation, and the Keiyo Line opened. The additional train kilometers alone reached a total of 90,000 km, a distance equivalent to that traveled by all of the trains operated by Seibu and Tokyu Railways combined.

JR East has also introduced 2,800 new EMUs. Although the company has added enough volume of trainsets to operate two private railway companies since it was formed, rush hour congestion has not noticeably improved.[22] Solving this problem continues to require a concerted effort by all JR East departments.

---

[21] As of 2016, there were approximately 16 million passengers per day using JR East for commuting to work or school in the Tokyo area, with revenues of approximately 800 billion yen.

[22] Rush hour congestion in the Tokyo area used to be 'about 180%' of capacity but has now improved to 'about 160%'.

9.10   Improving Station Images

The train station is the face each railway company shows the public; it can establish the corporate image. More than simply the place where passengers catch their trains, a station is the entrance to its city and a place where people mingle. In the magazine 'Le Moniteur universel' in an article entitled 'Railway Stations,' the 19th-century French poet Théophile Gautier discusses station architecture as follows.

> The train station is a nigh cathedral of humanity. People are attracted to it; countries mingle. Everything is there. It is the center of the ring of light emitted by railway lines stretching out to the ends of the earth. I may also add that the train station embodies a new style of architecture that people have sought but were unable to realize in the past, and it will continue to answer the newly evolving needs of the road. This was unable to exist in ages past. The fruits of architecture handed down from ancient temples and medieval churches have come to blossom in our train stations.

Traveling in Europe, one finds the splendid station architecture described by Gautier. Lyon Station in Paris, Victoria Station in London, Leipzig Station in Germany, and Milan Station in Italy are not more than just a gateway to the railway. They are urban monuments, as well as cenotaphs that teach us the history of architecture.

Railway stations are, however, more than legacies of the past. The original designs of certain new station architecture project our will for the future. Waterloo Station in London and the North Station in Paris, which were remodeled when the London-Paris Eurostar debuted, *Charles De Gaulle* Airport Station in Paris, the Satras Station in Lyon, and the soon-to-be-completed Frankfurt Airport Station are a few of the stations that illustrate that their new design radically changes the image of traditional station architecture.[23]

Most JNR station architecture has been long on function but short on originality. A typical example is a station called an over-track station. Passengers go up the stairs leading from the street or square to reach the ticket window and the entrance/exit with turnstiles. Functionally, this structure is adequate. It was convenient in that it allowed passengers to freely access the opposite side of the tracks. In terms of design, however, there were absolutely no distinguishing features among the approximately 300 such stations built throughout the country.

The *Shinkansen* seems to have been responsible for this functionalism in station design.

The *Shinkansen* concept itself was functional in nature. Station, railcars, and train schedules are all aimed at function and standardization. Although there was some criticism of the uniformity in station design, this was precisely the point of the *Shinkansen*. This design concept reflected the values of a period in which Japan continued to seek high economic growth with limited capital. In their day, *Shinkansen* stations were presented with the Architectural Institute of Japan Award.

JR East hoped to do something to change the image of its station. We did not, however, want to demolish Tokyo Station; it was the symbol of JR East. Quite a few

---

[23] Frankfurt Airport Station was completed in 1999.

people thought that the old JNR-era red brick structure should be torn down and a large-scale building constructed in its place. They justified this by pointing to the fact that 'the building was so old that repairs were difficult' and that 'architecturally speaking, the building was not an important building.' Granted, architectural value is a matter of subjective opinion, but no one could deny that, as a historical part of the urban landscape in Tokyo, Tokyo Station has left a strong impression on the people of Japan.

Taking a hint from the Orsay Museum in Paris, a station gallery was set up inside this station. Rather than entering into a debate about whether the current building should or should not be preserved, we felt that our effort was better spent creating the facts whose existence people would feel to be important. The completed gallery creates a unique space wonderfully set off by exposed brick (Fig. 9.9).

The twentieth century master painter Balthus honored us by saying that if he were to exhibit his paintings anywhere, he would choose this gallery. Over the 89 days of the Balthus exhibit in 1993, 100,000 people visited the Station Gallery.

To digress a bit, I would like to shed some light on the confusion surrounding the design of Tokyo Station. There is no basis for the popular idea that this station was modeled after Amsterdam Central Station in Holland.

In recognition of Dr. Kingo Tatsuno, the Tokyo Station architect, Tokyo Station Gallery sponsored an exhibition entitled 'Tokyo Station and Kingo Tatsuno' in 1990. In a section of the exhibit catalogue entitled 'Two Central Stations: Amsterdam and

Fig. 9.9 An exibition room of the Tokyo Station Gallery. *Photo* provided by the Tokyo Station gallery

Tokyo,' Dutch architecture historian Aart Oxenaar describes the Amsterdam Central
Station design created by architect Pierre Cuypers in the following manner.

> This extremely distinctive structure integrates a tradition of station building architecture that
> adopts Renaissance and Baroque palace architecture with Cuypers' own architectural and
> urban vision.

Cuypers is said to have designed this station as a piece of modern architecture
against a backdrop of the medieval Gothic architecture he so highly regarded.

Oxenaar then goes on to say, "Tokyo Station and Amsterdam Station obviously
bear extremely little resemblance to one another in terms of decorative elements.
Although I will not go so far as to say that the two stations have nothing in common in
terms of structure and materials, I find nothing to indicate that Tatsuno was influenced
by Cuypers or that railway technicians from Holland and Japan interacted in any
way. There is no proof that Amsterdam Station had architecturally influenced Tokyo
Station" (Figs. 9.10 and 9.11).

Japanese architect Takeo Amito explains the origin of this popular notion. "The
Amsterdam story can be traced back to around the end of the war. In post-war
architecture, domed roofs disappeared to be replaced by mountain-shaped roofs with
straighter lines. It is these straight lines that resemble the design of Amsterdam
Station. Also, Tokyo Station is not actually a red brick building. It is true that the
frame is made of brick, but the exterior walls are made of brick tiles".

The Tokyo Station building used today was completed in 1914. Before then,
the Tokaido Line terminated at Shinbashi Station in present-day Shiodome. Only

Fig. 9.10 Tokyo Staion known for its distinctive red bricks. *Photo* provided by Kotsu-Shinbunsha

Fig. 9.11 Holland's Amsterdam Central Station, a long, narrow building on the bank of a canal. *Photo* provided by the Railway Museum

the Yamanote Line was extended as far as the Gofukubashi, a temporary station constructed a bit north of Tokyo Station.

In 1909, five years before Tokyo Station was completed, EMU trains replaced those hauled by steam locomotives on the Yamanote Line. The typical circuits run by today's trains were introduced in 1925. Against this backdrop, Tokyo Station has not only come to symbolize the city of Tokyo but also stands as a witness to the history of Tokyo's urban transportation. This station is undoubtedly a grand monument to the Japanese railway.

If Tokyo Station couldn't be rebuilt, we felt strongly that a new building should be constructed in another location to symbolize the newly formed JR East. Ueno Station was nominated as an appropriate site. Since a large-scale redevelopment plan was already in place, building a skyscraper in this area seemed reasonable. We hoped, if possible, to create an original architectural design that would be appreciated by future generations. Unfortunately, we encountered delay after delay involving problems with the building blocking the sun in Ueno Park, some local opposition, and studies regarding the contents and economic feasibility of the building projects. During this time, Japan was hit with a recession and there seemed to be no hope that this project would ever get underway.

JR East also studied the possibility of remodeling Ochanomizu Station. The platform of this station is narrow, but it cannot be widened since it is located between a cliff and the Kanda River. Plans were therefore developed for the construction of a low building covering the tracks to house a passenger concourse. We sponsored a competition calling for proposals from professional architects. It followed the examples of the design competitions held to choose the designs of the Termini Station in Rome and the Milan Central Station. Although we received 252 applications and chose the most exuberant designs, this project never saw the light of day. When the remodeling was attempted, our engineers discovered that working on the Kanda River precipice to lay the foundations would be extremely difficult and incur enormous expenses. Like Ueno Station, this dream also burst like a bubble.

On a more positive note, JR East was steadily modifying its stations, albeit in more subtle ways. In opposition to tradition, unique elements have been added to a number of smaller station buildings. These designs include a series: the unique stacked umbrella-shaped roof of Iwaki-Hanawa Station (Suigun Line), the castle style of Funaoka Station (Tohoku main line), the two enormous tube structures laid side by side at Yabuki Station, the complete transformation of traditional station architecture at Akayu Station (Tohoku main line), and the carved steps leading to the ocean at Kuwakawa Station (Uetsu main line). Stations that would never have come into being during the JNR era, when stations were merely functional and uniform in style, were born one after another (Fig. 9.12).

In addition to spicing up the design, we have also considerably transformed station functionality. In this day and age of private cars, stations serving merely as sites for passengers boarding trains will not attract people. Fewer people at the station not only adversely affect the stations themselves but also lead to the decline of the area in front of the station. It is fine if people do not actually board the trains. We want the

Fig. 9.12 Unique station building, Akayu station. *Photo* provided by the Kotsu Shinbunsha

train station to serve as a place for people to get together. This concept is the basis of such experiments as the hot springs at Hottoyuda Station (Kitagami Line) and Takahata Station (Ou main line), the ski lift departing directly from Gala-Yuzawa Station (Joetsu *Shinkansen*), and the regional community centers equipped with a local archive library, galleries and other amenities at Innai Station (Ou main line), Naruko Onsen Station (East Rikuu line) and Yamagatajuku Station (Suigun Line).

We are also seeing gradual changes at larger stations. In addition to the Station Gallery, Tokyo Station platform design has also changed significantly. The areas beneath Tokyo Station have changed, as well. Anyone comparing the Sobu-Yokosuka Line area designed during JNR era and the Keiyo Line area designed by JR will appreciate the difference in design philosophy.

With a total of 3.28 million passengers a day on JR, private railway and subway companies, Shinjuku Station, the busiest station in the world, is a model of guidance design.[24] A new shopping corner has filled the empty space in Ueno Station, resulting from the noticeable decline in *Shinkansen* transfers. Ebisu Station has been remodeled into the glamorous Saikyo Line terminal and the stations at Nagano, Akita, Yamagata, and other places have also been renovated.

Less glamorous, but nonetheless important, is the issue of furnishing escalators. Most Tokyo area stations will be equipped with escalators within a few years.[25] Mobility with the least resistance is at the core of train station issues. In this vein, another important matter confronting the railways is future coexistence with road vehicles, particularly private cars. Today's car culture enables people to easily reach any destination, and unless stations provide facilities for drivers to park their cars without hassle, their functionality will be diminished by half. The Tohoku *Shinkansen* Kurikoma Kogen Station was recently constructed on land surrounded by rice paddies; its large parking lot, however, has brought more than twice as many passengers than predicted to this station. Annakaharuna and Sakudaira Stations on the recently introduced Nagano *Shinkansen*[26] also plan to provide similar facilities. Railway and automobile intermodality will be a critical issue for railway stations to deal with.

---

[24] As of 2019, Shinjuku Station had 3.53 million passengers per day.

[25] About 95% of stations with more than 3,000 passengers per day have escalators, elevators, etc., installed.

[26] This line was partially opened in 1997 between Tokyo and Nagano as the Nagano *Shinkansen*. In 2015, the Nagano–Kanazawa section opened and is now called the Hokuriku *Shinkansen*.

Chapter 10
Next-Generation Railways

In this chapter, various *Shinkansen* developments, which have different dimensions from the technological progress that resulted in the increase in maximum speed, will be presented. The author's perspective is a balance between high performance and cost, the bold challenge of new technologies and timid vigilance toward safety, and the provision of new services that will delight passengers.

As examples of how this has been realized, he introduces the mini-*Shinkansen* project, which realized the needs of residents along lines at a small construction cost and the birth of new demands such as *Shinkansen* commuting and the establishment of stations adjacent to airports. As the accident at Eschede Station in Germany shows, it is difficult to maintain the safety of railways, but as an expert in this field since his youth, he shares his knowledge that is difficult to hear elsewhere.

10.1   Mini-*Shinkansen* and Super Express

Soon after the JNR reform, I had the opportunity to talk with the chairman of a major computer manufacturer. We happened to get onto the subject of the exodus of factories from Tokyo to rural areas.

When I asked if his company also planned to move its Tokyo plant to the countryside, he replied,

> Of course we do, and do you know which prefecture most companies are choosing for their plants?
>
> No, which one?
>
> Niigata Prefecture
>
> Why Niigata?
>
> It's a natural choice, isn't it? What with the *Shinkansen* and the highway.

© Japan Railway Technical Service 2024
S. Yamanouchi, *If there were no Shinkansen*,
https://doi.org/10.1007/978-981-99-8890-7_10

It is true that new factories spring up and cities themselves change rapidly when the *Shinkansen* service is introduced. *Shinkansen's* service has completely changed the city itself of both Sendai and Morioka. Houses have also dramatically increased near the Kitakami Station area. While some believe that the *Shinkansen* service will bring development to rural cities, others are of the opinion that it will only further the concentration seen in Tokyo. While both claims are partly true, cities located along *Shinkansen* lines will in any event be significantly altered.

Five years after the Tokaido *Shinkansen* service was introduced, the next step in our plans, the construction of the Sanyo *Shinkansen*, began. *Shinkansen's* development, however, did not stop there. After the success of the Tokaido, suddenly prefectures all over Japan were calling for their own *Shinkansen*. Enacted in 1969, the New National Comprehensive Development Plan advocated *Shinkansen* lines throughout the country from as far north as Asahikawa in Hokkaido to as far south as Kagoshima in Kyushu by 1985.

Plans on this large scale would be unthinkable today, but at the time, the residue of rapid economic growth remained. Many people in Japan honestly believed that this could be accomplished.

In 1970, the national *Shinkansen* Railway Development Act was enacted as the basic regulation for *Shinkansen* construction procedures.

This law stipulated the procedures for building a *Shinkansen*. The Minister of Transport first drew up a basic plan outlining the location of the future *Shinkansen*. Lines not included in this plan will not be considered for *Shinkansen*. Inclusion in the Minister's plan meant making it into the big league. Up until 1973, nearly every line proposed by the New National Comprehensive Development Plan was included in the basic plan. In essence, the government officially sanctioned the 7,000 km of Basic Plan *Shinkansen* Lines.

Construction work did not, however, begin when a line was listed in this plan. 'Implementation plans' for projected lines should then be drawn up. These plans represented concrete construction plans concerning where the new *Shinkansen* would be constructed, and stations built. Implementation plans cannot be made by local governments on their own. Under the direction of the Minister of Transport, Japan Railway Construction Public Corporation develops these plans. So far, we can finally say we are benched.

Before construction could begin, however, there was one final step. Japan Railway Construction Public Corporation and other organizations that would be in charge of the actual construction were to develop a Construction Implementation Plan, which required the approval of the Minister of Transport. When this last plan was approved, we were finally up to bat.

All *Shinkansen* up through the Tohoku and Joetsu were constructed under these procedures. Despite receiving approval, however, the construction of the Narita *Shinkansen* remained unfinished because of fierce objections from the inhabitants along this new line. Construction came to a standstill. It entered the batter's box but struck out looking away.

Although the projected plan was set at the time of the reform of JNR and plans for projected lines for five other *Shinkansen* lines had been approved, the standstill prevented construction from starting. These five *Shinkansen*—the Hokkaido Line between Aomori and Sapporo, the Tohoku Line between Morioka and Aomori, the Hokuriku Line between Takasaki and Osaka via the Hokuriku region, and the Kyushu Lines between Fukuoka and Kagoshima and Fukuoka and Nagasaki—were dubbed the five projected *Shinkansen* lines.

Although local residents longed for some move that would allow construction to begin, JNR was in financial turmoil. There were no resources to spare for the *Shinkansen*. The government froze construction work on these projected *Shinkansen* lines in 1982.

The freeze was lifted immediately before privatization, and the reform of JNR was set in motion with a Projected *Shinkansen* Line time bomb ready to explode.[1]

In 1988, the government and the ruling parties formed the Projected *Shinkansen* Lines Advancement Exploratory Committee to give priority to some projected lines for construction and also formulate specific construction plans for these lines at the same time.

The committee's specific plan was, so to speak, a speckled *Shinkansen* construction plan. The construction of new high-speed lines was to be concentrated in only one part of the approved lines. These plans canceled the full-sized Shinkansen lines on some parts of the lines. Instead, a 'mini-*Shinkansen*' line was to be constructed by modifying the conventional meter gauge lines in that section to standard gauge.

Shinkansen-type infrastructure would be constructed for the Hokuriku and Kyushu lines. Initially, however, meter gauge tracks would be laid and limited express trains running on conventional meter gauge lines would be operated at high speeds on these new lines. Since the new infrastructure was to be constructed in isolated sections separated from existing *Shinkansen* lines, these changes were necessary. When the construction of this missing link was finally completed, these lines would be reformed to full-sized standard gauge *Shinkansen* lines. This type of line was called 'super express.'

Speckled *Shinkansen* construction would greatly reduce overall construction expenses, making the construction of projected *Shinkansen* lines once again feasible.

The speckled plan, however, generated fierce opposition from local residents. They complained that they had ordered eel, but loach and catfish were served instead. By loach, they meant the mini-*Shinkansen*, while catfish referred to the super express.[1]

Despite the significantly lower construction costs, the issue of who would pay for the projected *Shinkansen* construction cost was still a weighty problem. Privately-operated JR companies would not undertake unprofitable projects such as new *Shinkansen* lines. The JNR privatization had the effect of enabling it possible to make these points clearly.

Ultimately, JR agreed to finance construction only to the extent that it would remain profitable for them (referred to as the benefit limit). We also agreed to make yearly payments in usage fees for the new infrastructure after these lines were completed. The central and local governments would pay the remainder. Another

---

[1] Eel is a popular, fairly expensive dish in Japan. The loach fish has a similar shape but is smaller than the eel. Catfish has a similar feel but is an entirely different shape.

stipulation was that JR would not operate the conventional lines paralleling the new *Shinkansen*, which would inevitably operate at a loss once the new service was introduced. Only under these conditions would JR bear a portion of the construction expense.

The first projected *Shinkansen* line built was the Hokuriku *Shinkansen*, or so-called Nagano *Shinkansen*, between Takasaki and Nagano. Construction began in August 1989 and was completed in October 1997. Although initial plans called for a loach between Karuizawa and Nagano, this line was transformed finally into a magnificent eel. Ultimately, JR East bore one-third of the 840 billion yen overall construction expenses. After this, construction began on the Tohoku *Shinkansen* between Morioka and Aomori, as well as on other sections of the Hokuriku and Kyushu Lines.[2]

At the time of privatization, JR did not take possession of the *Shinkansen* infrastructure. Instead, it was leased from the *Shinkansen* Holding Organization, a government-operated entity.

But JRs bought the entire infrastructure of the existing *Shinkansen* in 1991. JR paid a total of 9.1 trillion yen, 1.1 trillion yen more than the officially estimated value. The extra capital would fund the future construction of the projected Shinkansen lines. In this way, JR continued to assist projected *Shinkansen* lines from its position as a group of private companies, as well.

Although the mini-*Shinkansen* disappeared almost completely from the projected *Shinkansen* lines, it popped up in an entirely different place. The first mini-*Shinkansen* was introduced between Fukushima and Yamagata in July 1992 with through *Tsubasa* service from Tokyo to Yamagata (Fig. 10.1).

This new idea at first came to my mind in the winter of 1983 when skiers flocked to the Joetsu *Shinkansen* for the first winter of passenger service. On Sunday evenings, the crowds were so thick on trains leaving Echigo-Yuzawa Station that the doors would not close, and station staff was forced to push passengers onto the trains from behind. It was like a scene from a rush hour on the Yamanote line. The line was still at the Omiya opening stage.

The *Shinkansen*, we realized, had the power to generate this much excitement, bringing about a substantial increase of passengers for railways. I had lived in Kaminoyama-Onsen, Yamagata Prefecture for a year when my classmates and I were evacuated from Tokyo during the war. I had also been skiing at Zao, Yamagata before I was married. It seemed reasonable that a *Shinkansen* running to Yamagata, one of the most famous ski resort areas in Japan, would attract a huge number of people. We then set about trying to figure out how to build a *Shinkansen* line to Yamagata.

The most difficult question was the projected *Shinkansen* line list.

Approved plans for projected lines were already in place for construction, and bypassing this list to begin *Shinkansen* construction on an entirely new section would be nearly impossible. Who knew when their five lines would be completed? It could take fifty years, and the railway line in this region could have disappeared by then.

---

[2] These lines have already been completed.

Fig. 10.1 *Tsubasa*, the first mini-*Shinkansen* to Yamagata. *Photo* Provided by the Railway Museum

I took down the statute books to study the national *Shinkansen* Railway Development Act provisions. Provision Two read,

"The term *Shinkansen* will refer to trunk lines accommodating high-speed trains running at 200 km/h or higher over major sections of track." A railway line therefore built on the same standard gauge tracks as *Shinkansen*, but with trains operating at less than 200 km/h would not be considered as *Shinkansen*. Thinking about it in this light, I realized that some private railway companies, Keihin Kyuko or Hankyu Dentetsu, are operating their trains on standard gauge lines as with the *Shinkansen*, but they were not a part of the *Shinkansen*. *Shinkansen* EMU trains could operate on conventional lines converted from meter gauge to standard gauge. Although operating at a much slower speed, these trains would definitely work for our purposes. The small tunnels and bridges on the conventional lines would not accommodate large *Shinkansen* trains, but compact trains could be developed especially for these lines. France's extension of its Paris-Lyon TGV trains onto conventional lines to Grenoble, Geneva, and other cities would also provide a useful example in this respect.

When I timidly pitched this idea inside JNR, not a single person voiced their support.

Everyone seemed to look at me as if to say, "You certainly have strange ideas" and "Don't you know that would only get in the way of projected *Shinkansen* line construction?".

I brought the idea to Tadatoshi Ino, the director general of the JNR Tokyo Metropolitan Area Headquarters, who happened to have graduated from the same primary school as I had but a few years ahead of me.

He thought it was interesting and suggested that we work on it together. We decided to look into Senzan Line between Sendai and Yamagata and Tazawako/Ouu Line between Morioka and Akita.

We had gathered some fairly specific data but still had no idea how to get the line constructed until I happened to tell a powerful local politician what we were working on. He was extremely enthusiastic and managed to push construction through.

Although we called this line the 'Yamagata *Shinkansen*,' this was only a nickname. The mini-*Shinkansen* does not officially fall into this category of *Shinkansen*. Even so, by introducing this new service of Yamagata *Shinkansen*, we were able to shave off about 40 minutes from the trip between Tokyo and Yamagata, which now takes two hours and 40 minutes. The greatest effect of the mini-*Shinkansen*, I think, was not the increase in train speed but rather the fact that people became more aware of the existence of the railway service itself.

Before this line was completed, someone suggested that those of us who had been evacuated to Kaminoyama-Onsen in Yamagata during our schooldays should take a reunion trip back there.

As we were making plans, one of my former classmates complained, "The *Shinkansen* makes it easy to get to Fukushima, but from there, we can't find how to get there!" Then I told them the limited express *Tsubasa* is running to Yamagata, connecting to the *Shinkansen*. They said they had never heard of it.

Almost all of my classmates were businessmen at the vanguard of their fields.

That evening, I asked my family if they knew that the limited express train *Tsubasa* was running to Yamagata. They replied coolly, "How on earth should we know that?".

Since those of us in the railway community know its network and train services so intimately, we tend to expect our passengers to at least know the railway lines and the limited express trains that are running on our lines. With more and more people traveling by car and a society in which we are bombarded with so much travel information, we tend to retain only the information that is of most interest to us personally. When I realized most people living in Tokyo were becoming less and less aware of lines other than the *Shinkansen*, it was a moment filled with discovery, as well as self-criticism.

The mini-*Shinkansen* served us best, I believe, by making people aware of the fact that there was now direct train service all the way to Yamagata. In fact, an executive of Yamagata Prefecture told me, "We are grateful that the Yamagata *Shinkansen Tsubasa* is announced hourly now at Tokyo Station. We have the *Tsubasa* train to thank for this".

Before the mini-*Shinkansen* could be developed, however, we faced a number of technical problems. Would we be able to develop bogies that would operate smoothly on both a high-speed *Shinkansen* line and a conventional line with many sharp curves? Would we actually be able to couple and uncouple trainsets for the first time in the history of the *Shinkansen*, during an intermediate station stop? To overcome the latter problem, we equipped the trains with a device that uses laser beams to accurately measure the distance between the cars to be coupled and display this information in the driver's cab.

The compact cars of the mini-*Shinkansen* would also leave wide gaps between the trainset and the platform at *Shinkansen* stations. Fencings were added to the platforms and folding steps to the cars for safety. These steps would automatically fold out when the train stopped at a station and fold in again once the train had departed and reached a certain speed.

There was, however, one incident that a passenger fascinated by the coupling work at the station missed his train's departure and, ended up on the step of the moving train. Fortunately, the driver made an emergency stop, and no one was harmed.

Safety devices presented another technical problem. Since *Shinkansen* lines use ATC devices and mini-*Shinkansen* lines use the ATS, trains had to be designed with both devices that can be switched from one to the other at Fukushima Station.

The suggestion was also made to lay three rails for the Yamagata mini-*Shinkansen* to allow meter gauge trains to operate on these lines as well. We wanted, however, to avoid this at all costs. Mixed gauge lines involve extremely complicated turnouts and track devices. This section of the track is subject to considerable snowfall in winter, which could cover the turnouts and stop operations. In my many years of experience in railways, complicated structures are certain to experience frequent malfunctions and failures. Simple is best, as they say.

Three rails would present another problem, as well. The extra rail would increase the expansion and contraction caused by the difference in summer and winter temperatures, which would subject the tracks to considerable stress.

Compared to a full-sized *Shinkansen*, the slower mini-*Shinkansen* speed meant a 20 minutes longer trip between Fukushima and Yamagata, but only one-twentieth of the construction expense. The gauge conversion of a conventional line also almost completely eliminated the need to purchase land, and we weren't faced with the annoying problem of what to do with a parallel conventional line. Trains can also stop freely at intermediate stations. A full-sized *Shinkansen* would have meant only one stop on the new line.

Most beneficial of all, these lines could be constructed quickly. The Akita mini-*Shinkansen* was the next line to be built, followed by the extension of the Yamagata *Shinkansen* to Shinjo. Seeing our success with these lines, I got the impression that other regions of Japan would have areas where loaches can live. We hoped the Yamagata mini-*Shinkansen* would change rural residents' opinion about the *Shinkansen*. I believe that we were able to so easily introduce the mini-*Shinkansen*, because it fulfilled the expectations of a certain segment of the population. But it wasn't so. People who once order eel will be satisfied with nothing else.

10.2   TGV Sets 515 km/h Record

On May 18, 1990, France set a remarkable speed record. During a series of high-speed test runs on the TGV Atlantic Line, which SNCF would soon be opening, a special test run train achieved a record speed of 515 km/h.[3] Two years before, an ICE train in Germany had reached 410 km/h for the first time in the history of the railway. France's new record, however, broke through this by an incredible margin.

Having set a previous record of 331 km/h, SNCF began working toward this new record to prove to the world its technological power. Speed may not be the only factor in technological capability, but there is no doubt that this record is impressive (Fig. 10.2).

JR East achieved a speed of 425 km/h three years later.[4] The lack of power generated by EMU trains and the absence of a sufficient length of straight sections on which trains can run at high speeds make it difficult for us to achieve anything higher.

In the case of locomotive-hauled trains like France's TGV using power cars, test run trains can be configured for extraordinary power per ton by reducing the number of passenger coaches coupled between the power cars. A particularly powerful electric motor was equipped in the power cars on the Atlantic Line test run trains, and the

Fig. 10.2 French TGV set a world speed record of 513.3 km/h at that time. *Photo* Provided by Keiji Musha

[3] The TGV reached a maximum speed of 574.8 km/h in 2007.

[4] As mentioned in Chap. 8, three years later, in 1996, JR Central's Type 955 (prototype) reached a maximum speed of 443 km/h.

voltage of the electric current was set higher than normal. Larger high-speed wheels also replaced the usual wheels. The original 10-passenger coach formation of the Atlantic Line TGV was modified to only four coaches coupled between the power cars (Table 10.1).

It was not impossible to replace JR East's *STAR 21* electric motor with a particularly powerful one, but we did not feel the need to go to these lengths. High-speed test runs for *STAR 21* were conducted between Nagaoka and Niigata, where there is a relatively sharp curve near Nagaoka Station. Even when we managed to increase the speed at which trains ran on this section to 250 km/h, the maximum speed we were able to reach overall was 425 km/h.

I myself was present at one of the test runs and watched the needle on the speedometer refuse to move once it reached 400 km/h. Light rain was causing the wheels to slip, and small plastic beads were scattered on the rails as the train accelerated. Even under these conditions, the train was extremely stable, and we could have attempted higher speeds if the *STAR 21* had been more powerful and the line had had longer sections of straight track.

Table 10.1 Comparison of high-speed rolling stock features

Country		France			Germany		China	Japan	
Car type		TGV-A	TGV-R	TGV-Duplex	ICE-3 (403)	Velaro D (407)	CR400AF CR400BF	N700	E5
Maximum operation speed (Km/h)	a	300	320	320	320	320	350	300	320
Number of cars (Notes)	b	12 (2P + 10 T)	10 (2P + 8 T)	10 (2P + 8 T)	8 (4 M + 4 T)	8 (4 M + 4 T)	8 (4 M + 4 T)	16 (14 M + 2 T)	10 (8 M + 2 T)
Capacity (pax.)	c	464	360	510	429	426	556	1323	731
Output (KW)	d	8,800	8,800	8,800	8,000	8,000	9,750	17,080	9,600
Weight with full of passengers	e	484	416	424	440	N.A	N.A	700	496
Length (m)	f	238	200	200	200	200	209	405	253
Capacity to length (pax./m)	c/f	1.9	1.8	2.6	2.1	2.1	2.7	3.3	2.9
Output/weight (KW/t)	d/e	18.2	21.2	20.8	18.2	N.A	N.A	24.4	19.4
Year in service	g	1989	1993	1996	2000	2013	2017	2007	2011

Notes; P = Power car, T = Trailer car, M = Motor car
Source; High-Speed Railways of the World JARTS 2023

Spending enormous amounts of money on a race for speed simply to save face seems childish. It would be impossible to challenge a speed of 515 km/h on Japan's *Shinkansen* tracks, which are full of curves and tunnels.

Still, the French speed record had a significant impact on Japan, and the question of just how fast trains could run became a popular topic of conversation. During the JNR era, the Railway Technical Research Institute announced that its research results indicated a maximum possible speed of 350 km/h, as long as steel wheels and rails were used. In light of France's accomplishment, people wondered whether this research had been flawed.

So, I ordered the research report from that time.

Dr. Tadashi Matsudaira, then director general of the Railway Technical Research Institute, published a report entitled 'Japan's Future Ultra-high Speed Railway' in the January 1978 volume of the Japan Society of Mechanical Engineers magazine. He stated that rolling stock vibration would aggravate stability once trains reached speeds of more than 350 km/h, then went on to explain why.

One major obstacle we now face is the problem of adhesion. The propulsion of trains currently entails the use of vehicle-borne motors, a motor to rotate the wheels, and the frictional force, namely adhesion, working between the wheels and the rails. In opposition to the gradual decline in adhesion as train speed increases, the train's running resistance rapidly increases as speed increases. At a certain speed, these two elements coincide. At higher speeds, propulsion no longer prevails over resistance, and the wheels will only slip no matter how much vehicle-borne horsepower is increased. Under these conditions, trains will not speed up. From *Shinkansen* EMU's research data, we can infer that, under the worst-case scenario of wet rails, the maximum speed would be approximately 350 km/h.

Apart from the problem of rolling stock vibration, please focus your attention on the last part of this excerpt.

Dr. Matsudaira asserts the need to take worst-case scenarios of rain and other problems into consideration when dealing with trains that operate 365 days a year. In such cases, trains would be limited to approximately 350 km/h.

Daily operations are, of course, completely different from a single round of high-speed test runs. For the French TGV test run, extremely powerful power cars were used, and track and overhead contact wire were carefully prepared for high speed. The test was realized on a gently down-sloped section on a sunny day. While most definitely superb, these results are not likely to be achieved under everyday railway operations. A successful test run is entirely different from a daily commercial service.

So many experiences provide plenty of examples of successfully tested technologies that failed in practical application. In the railway business, extremely long periods of time between successful tests and the practical application of this technology are particularly common.

During test run of a high-speed EMU train speeds exceeding 200 km/h took place in 1903 in the suburbs of Berlin. Sixty-one years later, the Tokaido *Shinkansen* was the first service train to actually operate at this speed. SNCF set a record speed of

331 km/h during test runs in 1955, but we have yet to see a service train operating at this speed.[5]

10.3 German ICE Derails

Taking a variety of issues into consideration, with current environmental problems at the national level in Japan, the issue of micro-pressure waves generated when high-speed trains enter tunnels, as well as future energy problems, global environmental problems, and cost–benefit issues, it would appear that the Japanese *Shinkansen* speed is reaching its limit. We seem to have reached an age when, rather than increase the speed of trains, future *Shinkansen* technology will be focused on other purposes.

Two accidents in 1998 had extremely significant implications for the *Shinkansen*.

The first was the ICE derailment at Germany's Eschede Station on June 3, 1998. Not only did this accident illustrate the horror of high-speed railway accidents, but it was also the latest sign pointing to safety as the most important issue for high-speed railways.

Although the precise cause of this accident is not as yet clear, a wheel on the second passenger car from the front was apparently cracked. After running on the rails for several kilometers without derailment, the car derailed at the station turnout. Its wild swaying caused the car derailment to crash into the road bridge pier at the exit of the station and the following 12 railcars to pile one after the other on top of the derailed car. It was a terrible tragedy (Fig. 10.3).

After the accident, Mr. Heinisch, the chief technical officer of DB, sent me two letters.

In a letter dated June 24, he wrote: "We are still reeling from the shock of the accident at the Eschede Station that killed 100 people. According to the latest inspection results, the direct cause of the accident was the destruction from the inside of the wheel tire with the rubber, called Type 64 wheels, used by the first generation of ICE cars, which were so thin that they were near the limit of their preventive replacement. It remains to be seen whether this failure was due to a defect in the material or to the synergistic effect of the thinner wheel thickness and some mechanical phenomenon".

I also sent a polite letter of sympathy and received a second letter dated August 4.

In this letter, he suggests the following three points:

1. The rubber between the wheels and tires of this train is a structure that was determined after careful study, the wheels and tires are held together by fixed rings, and no heat treatment such as shrink fit is used.
2. In adopting this wheel structure, a long bending endurance test was conducted, and it is not yet known if the damage was caused by a defect in the material or if the problem was due to the excessive load applied.

---

[5] The E5 series in Japan and the TGV in France both operate at 320 km/h, but have not exceeded 331 km/h. However, according to UIC data, some lines in China, including the Beijing–Tianjin route, provide a train operating service at 350 km/h.

Fig. 10.3 On June 3, 1998,
an ICE train derailed and
overturned at Eschede
Station in northern Germany.
Photo Provided by Tokyo
Shinbunsha

3. The first generation ICEs used rubber sandwiched wheels for ride comfort, and as a result, ultrasonic inspections also concentrated on the presence of abnormalities on the tire surface and did not focus on inspecting the inside of the tire.

He added, "We would like to use this accident as an opportunity to re-examine the entire safety system, including the interrelationship between humans and machines".

These first-generation ICE train wheels have slightly different designs than the wheels used on Japanese *Shinkansen* and French TGV trains. *Shinkansen* wheels are simply fitted to axles, while the German ICE wheels are triple-layered, with a tire fitted onto the outside of the wheels. In the past, most wheels on Japanese cars were also manufactured using this triple structure. The part of the wheel coming into direct contact with the rails is subject to increased wear. Wheels are covered with tires made of hard material so that only the outer layer needs to be replaced when worn. In terms of material expenses, replacing the entire wheel is more costly than this partial replacement method.

There is the potential for these tires to come loose, albeit rarely, and checking for loose tires was the most crucial factor in rolling stock inspection. When I was the head of a passenger car depot, most passenger car still had these tires fitted to the wheels. To make the inspector's job easier, the tires and wheels were marked with white paint. When inspectors saw that this mark was out of alignment, they knew that the tire was coming loose.

Around the time that the *Shinkansen* was being planned, an integrated wheel, which integrates the tire and wheel as one unit, was developed. From the outset, *Shinkansen* railcars have been equipped with integrated wheels.

Why, then did Germany still use wheels fitted with tires? The ICE trains did not, in fact, use the older structure wheels. It was a kind of new challenge.

ICE trains had suffered from a strange phenomenon in which circular wheels warped into an oval shape as the trains traveled. The reason for this is not absolutely clear, but according to Mr. Heinisch, it was caused by the hardening of track ballast. Oval wheels naturally made riding quality extremely uncomfortable. To solve this problem, Germany developed a type of "elastic wheel" with rubber sandwiched between the wheel and tire. DB equipped rolling stock depots for ICE trains with the ultrasonic equipment needed for inspections of scratches on the wheel surface.

Not only did elastic wheels prevent wheel deformation, but they also effectively reduced the noise generated by a train running. The USA had adopted elastic wheels on its new streetcars called PCCs as far back as the 1950s, and these cars were extremely quiet. In Japan as well, Tokyo municipal trams had also tried these wheels on prototype tram cars. Many German streetcars are currently equipped with elastic wheels.

The possibility of adopting elastic wheels on the Japanese *Shinkansen* was seriously studied at one time as a noise-reduction measure. The idea was ultimately rejected due to safety concerns when using them at high speeds and the fact that the noise from the pantographs could be more serious than that generated by wheels when trains reached high speeds.

DB used elastic wheels only on its 60 first-generation ICE trainsets. Later trainsets were equipped with integrated wheels. This may have been because they also experienced problems with this type of wheel.

Criticizing DB after this accident for its choice does not seem absolutely justified. Prudent decision-making and a spirit of challenge are both necessary when introducing any new technology.

Before adopting the elastic wheels, DB conducted a series of careful tests. We also must keep in mind that there were no accidents for 10 years after the ICE prototype was first introduced. The nature of technology is such that it takes 10 years for certain types of accidents to occur, and the balance between cautiousness and challenges is a very delicate issue for safety and technological innovations.

This accident was preceded by a tramcar in Hanover that also suffered damage to its elastic wheels. In contrast, some have pointed out that the problem is that DB was not informed of the accident.

Thirteen days after the Eschede Accident, a wheel on a coach of a British high-speed IC225 express was damaged and derailed when the train was traveling 200 km/h north of London. Fortunately, the accident was not serious. The derailed coach was equipped with integrated, not elastic, wheels. The hole by which the wheel is lifted by a crane had apparently cracked, which proves that integrated wheels could also be subject to safety problems.

Just nine days after the Eschede Accident, DB found itself facing an accident once again. This time it was a collision close to Karlsruhe. As the saying goes, "Accidents bring more accidents." Although the timing of these accidents was coincidental, they were facing a substantial and extremely serious situation regarding their safety.

Since 1997 DB had been plagued by a series of major accidents, including a collision between passenger and freight trains, and a tank car explosion. None of these accidents was directly interlinked. Any series of accidents, however, should be taken as a safety warning requiring a reexamination of technology and the element of human error.

Nine days after the Eschede Accident, I had the opportunity to visit Germany. Upon arriving at the airport in Frankfurt, I immediately headed to the Hauptbahnhof to catch the 18:05 ICE *Limmat* train to Zurich.

However, the ICE that arrived about 30 minutes late was not the bright white-colored trainset with the red stripe on its body, but rather an old passenger trainset with only five cars, hauled by an old 103 class locomotive. Nevertheless, the second-class cars were quite crowded, as they were originally supposed to be a 12-car trainset, although four more cars were added in Frankfurt. After the accident, all first-generation ICEs, 60 trainsets, were taken out of service for inspection. This train was a hastily composed train with a collection of old passenger cars. The running speed was also much slower than the 250 km/h ICE, about 160 km/h.

All first-generation ICE wheels underwent conscientious and thorough inspections immediately after the Eschede Accident, which took many of these trains out of commission.

In the end, DB decided to replace all 7,000 elastic wheels on first-generation ICE trains with normal integrated wheels. Since service was first introduced on the Tokaido *Shinkansen*, the Japanese *Shinkansen* has transported a total of 15 billion passengers. The French TGV has seen 500 million passengers. German ICE trains have transported a total of 100 million passengers. Then this accident happened.

10.4   Heading for a New-Era's Railway

Nothing built by human beings is perfect, and we never claimed that we were setting out to create a 'legend of safety around the *Shinkansen*.' More than 30 years without a major accident is certainly an incredible feat. As time goes on, however, it will be even more difficult for the *Shinkansen* to live up to this legend.

Shinkansen development was based on a delicate balance of daring to adopt new technologies and a caution bordering on cowardice with regard to safety. While most of the problems that cropped up after passenger service was worked out and many improvements were made, safety was carefully maintained through daily inspections.

Hideo Shima, a former chief engineer and the father of the *Shinkansen*, passed away on May 18, 1998. We felt that day as if an era had come to end.

Shima said something that has stayed with me ever since. When we were approaching the introduction of Tohoku *Shinkansen* service, I asked whether he was nervous about the new line since it had taken a rush job to be readied in time. He replied,

I am not at all worried about the Tohoku *Shinkansen*. We simply have to follow the same steps we took with the Tokaido *Shinkansen*. What I worry about is how the Tokaido *Shinkansen* will be running tomorrow. That is an unknown field.

These words seemed to me to express an issue inherent in technology. The *Shinkansen* has operated for more than 30 years without a major accident. But we do not know what tomorrow will bring. We must always be careful; we must work boldly yet cautiously, taking the smallest sign of an abnormality seriously.

In the *Shinkansen*, Shima and his engineers created a splendid new railway concept. Now we must take his lead and form our own next concept for the next new railway system. Safety, environment, cost, information, appeal, and convenience will, I believe, all be key.

We are already moving in this direction. In the area of safety and information, we are now applying the latest information technology to develop new signal and rolling stock systems.

Although not directly related to safety, another innovative information system is 'COSMOS,' the new *Shinkansen* traffic control system introduced in 1998. In terms of rolling stock and infrastructure maintenance, new devices, instruments, and database management techniques are potential tools for advanced technology in maintenance, which in the past has largely depended on human experience and skill. Medical science is, in a sense, human maintenance. It should therefore be possible to incorporate technologies as advanced as those we see in medical science in the maintenance of inanimate objects, as well.

The mini-*Shinkansen* proposal was an innovative concept in the areas of cost and convenience. Future *Shinkansen* development may incorporate even simpler, lower-cost designs or even meter gauge *Shinkansen*. Lightweight rolling stock was not only a step toward lowering operation costs but also toward lessening the environmental impact of trains through energy conservation.

The *Shinkansen* began with the concept of simple infrastructure and operation on a simple timetable. This ingenuous train changed the way people lived. Traveling back and forth between Tokyo and Osaka in one day came to be common. University exam cram schools suddenly popped up near *Shinkansen* stations, and instructors began to use these trains to travel between classes at schools in different cities.

In preparation for the *Orient Express* trip through Japan a few years ago, I visited Tokuyama City. While I was there, I asked a young woman what she did on the weekend. She surprised me by replying, "I take the *Shinkansen* to Hakata to go out with friends".

I've also heard stories of people living in a snowy region of Tohoku who take the *Shinkansen* during the winter to play golf in the Tokyo suburban area when it's snowing at home (Fig. 10.4).

When the *Shinkansen* was first built, I doubt anyone would have predicted anything like passengers using these trains to commute to the office every day. Today, 36,000 people buy *Shinkansen* commuter passes to travel to work and school.[6] This is a ten-fold increase in the number of passes sold since the privatization of JNR.

---

[6] As of 2019, there were approximately 140,000 commuter pass users per day.

Fig. 10.4   Gala Yuzawa Station where visitors can enjoy a one-dat skiing trip from Tokyo. Utilizing a depot for track maintenance cars. In front of the station is a ski resort. *Photo* Provided by the Gala Yuzawa Company

As people's lifestyles change and society changes with the *Shinkansen*, it will no longer be possible to have the same simple railcars and one-pattern train schedules as 30 years ago. When the Tokaido *Shinkansen* was first established, not a single *Hikari* train stopped at Yokohama, but today many *Hikari* trains stop there.

With so many people owning their own cars, we must find a way for the *Shinkansen* to reach out to cars and airplanes. The changes taking place are eloquently expressed by the success we have had with the Kurikoma Kogen Station. The large parking lot has brought more than twice the estimated number of passengers to this station. Connection terminals for the French TGV and German high-speed lines are being built at major airports in these two countries, and intermodality with cars and airplanes will, I believe, be an important element of future *Shinkansen* concepts.

At the same time, the *Shinkansen* itself must constantly offer new attractive services. Passengers on the Akita and Nagano *Shinkansen* have increased by 30 to 40% despite the fact that the service was introduced in 1997 during a period of severe economic recession. As many as 200 passengers have bought commuter passes for the 200 km between Nagano and Tokyo, many passengers coming from Takasaki to Tokyo also prefer to take the new *Asama* train.

The *Shinkansen* creates a new lifestyle, and its passengers today are always looking for new services. Safety and the challenge of new technologies and services are the genesis of the *Shinkansen*.

Epilogue

On October 16, 1996, the leading French newspaper 'Le Monde' ran a full-page interview on the privatization of railways in Germany with Heinz Dürr the first chairman of the privatized German Railway Co (DB). At the end of the interview, Dürr said, "I look to the privatization of railways in Japan as a model. The Japanese system is perfect. Trains run on time, and productivity is high. And JR does not receive a single yen in government subsidies."

In my more than 40 years on the railway, I have experienced the heights of glory and the depths of hell. The *Shinkansen* was the pride of Japan and a Renaissance for the railways. Yet only 10 years later, the management of JNR was plunged into bedlam. These experiences offer considerable material for self-reflection and a myriad of lessons to be applied today when all of Japan seems to be suffering from the "JNR syndrome."

Neither the creation of the *Shinkansen* nor JNR reform could, I believe, have been possible at any time other than the moment they took place. At each of these moments, we were presented with a once-in-a-lifetime opportunity. I shudder to imagine what would have come to pass if we had not built the *Shinkansen* 30 years ago and restructured JNR 10 years ago. There is no doubt in my mind that Japan's railways would have faded into extinction.

They say the corporate lifespan is 30 years. Revolution has revived the Japanese railways twice; our experience and spirit will be our foundation for living up to the challenge of the next great railway innovation.

Shuichiro Yamanouchi.

November 1998.

© Japan Railway Technical Service 2024   241
S. Yamanouchi, *If there were no Shinkansen*,
https://doi.org/10.1007/978-981-99-8890-7

References

Title	Author(s)	Published by
Published in Japan		
The Centennial History of Japanese National Railways		JNR
Twenty Years of the Shinkansen		JNR Shinkansen Directorate General
Ten-Year History of the Shinkansen		JNR Shinkansen Directorate General
Outline of Deliberations of the Committee for the Investigation of Train Speed		JNR
Memorial Lecture Record; 50th Anniversary of the Railway Technical Research Institute of Japan		RTRI
Report of the Committee for Shinkansen Transport Disruption Countermeasures		JNR
Report of Motive Power Modernization Committee		JNR
Tokaido Shinkansen from its opening to stabilization		RTRI
Ten Years of Progress		RTRI
Research on High-Speed Railways		Railway Research-Culture Promotion Foundation
Searching the Origin		Transport Cooperation Association
Catalogue; Tokyo Station and Kingo Tatsuno		Tokyo Station Gallery
New Tracks -Structure and Maintenance	Masao Suda and others	Japan Railway Civil Engineering Association
Shinkansen and the Universe	Hideo Shima	Railway System Research
Encyclopedia		Heibonsha
Journal of the Japan Society of Mechanical Engineers		Japan Society of Mechanical Engineers
Railway Research Review		RTRI
Published outside Japan		
Les défis du TGV	Jean François Bazin	Denoël
Direttissima Italien	Werner Hardmeier, Ascanio Schneider	Orell Füssli

(continued)

(continued)

Title	Author(s)	Published by
Intercity	John Gough	IAN ALLAN
Histoire de la traction électrique	Fernand Nouvion and others	La Vie du Rail
Die Personenwagen der SBB		Eisenbahn-Amateur
La revue generale des chemins de fer		ESME
La Vie du Rail		La Vie du Rail